A Review of Cardiac Radiology

By

John H. Woodring, M.D.

Staff Radiologist
Radiology Service
Lexington Department of Veterans Affairs Medical Center
Lexington, Kentucky

and

Adjunct Professor
Department of Diagnostic Radiology
University of Kentucky Medical Center
Lexington, Kentucky

ISBN: 1-4107-4947-9 (e-book)
ISBN: 1-4107-4946-0 (Paperback)

Library of Congress Control Number: 2003093182

This book is printed on acid free paper.

Printed in the United States of America
Bloomington, IN

1stBooks – rev. 06/20/03

To my wife

Catherine Anne Martin, M.D.

Preface

This book has been designed as a reference for several different purposes.

1. For diagnostic radiology residents preparing for written and oral board examinations in cardiac radiology.
2. For board certified diagnostic radiologists preparing for recertifying examinations in cardiac radiology.
3. For young academic diagnostic radiologists given the challenge of teaching cardiac radiology to diagnostic radiology residents.

Several years ago I was helping a group of senior diagnostic radiology residents prepare for their oral board examinations in cardiac radiology. I started out by putting seven different chest radiographs on a panel of view boxes. The seven cases were: a normal chest radiograph, a case with an enlarged heart with normal pulmonary vascularity (pericardial effusion), a case with pulmonary venous hypertension (mitral stenosis), one with overcirculation of the pulmonary vascularity (renal failure), a case of pulmonary arterial hypertension, a case with decreased pulmonary vascularity (Ebstein's anomaly), and a case of systemic arterial collateral circulation to the lungs (pseudotruncus). I started to explain that by identifying the pulmonary vascular pattern, and by understanding a few key concepts concerning heart size, specific chamber enlargement, and other important factors such as mirror-image right aortic arch, one could usually arrive at a specific diagnosis or a reasonably short differential diagnostic list for any cardiac case. As I was explaining this, one of the residents stopped me and said, "So what you're trying to tell me is that there actually is a way to approach chest radiographs from the standpoint of cardiac disease?"

"Yes," I replied, but his statement was telling. Few diagnostic radiologists teach cardiac radiology and most practicing radiologists don't know an approach to cardiac radiology. As a result most diagnostic radiology residents are left to try to teach themselves enough cardiac radiology to guess their way through their oral board exams.

The situation for residents preparing for written boards is difficult as well. The information they are expected to know is spread throughout a huge volume of work consisting of classic textbooks and articles written over the last 50 years spanning the disciplines of anatomy, physiology, embryology, cardiology, thoracic surgery, pediatrics, and diagnostic radiology. Most residents simply cannot muster such a review while they are simultaneously preparing for questions on all the other disciplines of diagnostic radiology as well. Many frantically pour over banks of old test questions on cardiac radiology only to become extremely frustrated when they discover that they can't find the answers to many of the questions.

Although no book can be totally comprehensive, in this book I have covered most of the material residents are expected to know on their written and oral exams in cardiac radiology. Furthermore, I have outlined a relatively simple but organized approach to the interpretation of chest radiographs from the standpoint of cardiac disease along with advice about how to handle certain difficult cases or situations that may arise during the oral board examination. My hope is that this book will help diagnostic radiology residents teach themselves cardiac radiology. Similarly, this book could also aid board certified diagnostic radiologists who are preparing for their recertifying exams. Furthermore, this book could be useful to young academic diagnostic radiologists who are assigned to teach cardiac radiology to diagnostic radiology residents.

Table of contents

Chapter 1.

A systematic approach to plain film cardiac diagnosis

I. **Plain film approach to cardiac diagnosis**
   A. **Step 1 – determination of pulmonary vascular pattern**
      1. Always begin your approach to a case by determining the pulmonary vascular pattern. The pulmonary vascular pattern indicates what physiologic effect the cardiac abnormality is producing and allows you to narrow your differential diagnostic list to a reasonable few likely causes, or occasionally a single most likely cause. In many cases, the pulmonary vascular pattern is more useful than heart size and signs of chamber enlargement in arriving at a correct diagnosis.
      2. For example, left atrial enlargement occurs in both mitral stenosis and ventricular septal defect. However, the presence of redistribution of pulmonary blood flow from pulmonary venous hypertension would indicate mitral stenosis, whereas overcirculation of the pulmonary vascularity would indicate ventricular septal defect. Although this example may be somewhat oversimplified, it is useful in illustrating the point that a correct diagnosis may not be possible based upon evaluation of heart size and chamber enlargement alone. Remember that the "key" to plain film cardiac diagnosis is the pulmonary vascular pattern!
   B. **Step 2 – assessment of heart size and shape**
      1. Decide whether the cardiac silhouette is enlarged or of normal size
      2. Look for specific chamber enlargement
   C. **Step 3 – 4R and 5L analysis**
      1. The 4R and 5L analysis is a systematic approach to analyzing the heart and vascular pedicle for specific physiologic, anatomic, and pathologic clues to cardiac diagnosis. The 4R and 5L analysis is also helpful in determining the heart size and in detecting chamber enlargement. "R" stands for steps of the analysis concentrating on the right side of the cardiac silhouette and vascular pedicle, while "L" stands for those steps concentrating on the left side of the cardiac silhouette and vascular pedicle.
      2. By routinely performing the 4R and 5L analysis, you will discover a number of important clues to cardiac diagnosis that might escape your detection otherwise. After a while this process becomes automatic and requires only a few seconds to perform.
   D. **Step 4 – extracardiac analysis (evaluation of the rib cage, sternum, spine, and upper abdomen for clues to the diagnosis)**
      1. Down's syndrome, and therefore atrioventricular canal, may be associated with only 11 pairs of ribs.

2. Bilateral inferior rib notching is a clue to the diagnosis of coarctation of the aorta.
3. Straight back syndrome may be associated with mitral valve prolapse.
4. Scoliosis may be seen in Marfan's syndrome, tetralogy of Fallot, and truncus arteriosus.
5. Pectus excavatum may be seen in mitral valve prolapse and Marfan's syndrome.
6. Anterior bowing of the sternum may be seen in Marfan's syndrome and is suggestive of cystic medial necrosis of the ascending aorta.
7. Polysplenia and asplenia are associated with abnormalities of thoracic and abdominal situs.

E. **Step 5 – list differential diagnosis or provide specific diagnosis if possible**
1. Only after completely analyzing the radiograph in a systematic fashion, beginning with determination of the pulmonary vascular pattern, followed by overall assessment of cardiac size and shape, the 4R and 5L analysis, and finally the extracardiac analysis, should you attempt to make a diagnosis.
2. In many cases you will only be able to give a reasonable differential diagnosis; however, by combining the pulmonary vascular pattern with specific findings from the 4R and 5L analysis you will find that you may be able to give a specific single diagnosis.
3. In the remainder of this chapter we will concentrate on the 4R and 5L analysis. In chapters 2 through 8 we will discuss how combining the pulmonary vascular pattern with specific findings identified from the 4R and 5L analysis can be used to generate limited differential diagnostic sets or a specific diagnosis.

II. **There are 6 patterns of pulmonary vascularity.**
   A. Normal
   B. Pulmonary venous hypertension (redistribution or cephalization)
   C. Overcirculation (pulmonary vascular plethora)
   D. Pulmonary arterial hypertension
   E. Decreased vascularity
   F. Systemic arterial collateral circulation

III. **4R and 5L analysis**
   A. **R1 – right aortic arch**
      1. Normally the aortic arch is located to the left of the trachea. However, you may easily overlook a right aortic arch unless you specifically look for one. The purpose of this step is to force yourself to look for a right aortic arch as part of your routine approach to chest radiographs.
      2. May be an isolated anomaly
      3. Often associated with vascular rings
      4. Frequently a clue to associated cyanotic heart disease, especially if there is mirror-image branching

B. R2 – enlargement of the superior vena cava (SVC)
1. Enlargement of the SVC can be determined by assessing the midline-to-right (MR) distance and midline-to-left (ML) distance of the vascular pedicle.
   a. Draw a vertical line over the midline of the spine (the spinous processes) at the level of the aortic arch
   b. The MR distance is defined as the distance from the midline to the point where the SVC crosses the right upper lobe bronchus. The ML distance is the distance from the midline to the point where the left subclavian artery arises from the aortic arch.
   c. Normally the MR distance is less than the ML distance. Because the SVC is a thin-walled, flexible vein it will distend in response to increased pressure or systemic blood volume, whereas the left subclavian artery, being a thick-walled muscular artery, will not dilate acutely in response to increased pressure or systemic blood volume.
   d. If the MR distance is greater than the ML distance, the SVC is either enlarged or displaced to the right.
2. Causes of enlargement of the SVC
   a. Increased systemic blood volume (plasma volume overload)
      1) Iatrogenic
      2) Renal failure
   b. Increased pressure in the right atrium (RA)
      1) Right ventricular (RV) failure of any cause
      2) Tricuspid regurgitation
         a) Bacterial endocarditis with valvular vegetations
         b) Ebstein's anomaly
      3) Pericardial tamponade, constrictive pericarditis
   c. SVC thrombosis or occlusion
3. Enlargement of the SVC does not occur in patients that have only increased pulmonary blood flow from acyanotic left-to-right shunts. This is a distinguishing feature between overcirculation caused by plasma volume overload and that caused by congenital left-to-right shunts.

C. R3 – enlargement of the azygos vein
1. Normal size of the arch of the azygos vein
   a. The azygos vein should not measure more than 1 cm in diameter on an upright posteroanterior (PA) chest radiograph.
   b. The azygos vein should not measure more than 1.5 cm in diameter on a supine anteroposterior (AP) chest radiograph.
   c. There is considerable variation in the size of the azygos vein for individuals within the population as a whole. In many normals the azygos vein is much smaller than the upper limit of normal and comparison with baseline normal chest radiographs becomes crucial in assessing changes in size of the azygos vein. For example, if a patient's azygos vein is normally 4 mm in diameter on an upright PA chest radiograph, it is quite enlarged for that patient when it becomes

8 mm in diameter, even though the 8 mm diameter is within the range of normal size for the entire population.

2. Causes of enlargement of the azygos vein
   a. Increased systemic blood volume (plasma volume overload)
      1) Iatrogenic
      2) Renal failure
   b. Increased pressure in the RA
      1) RV failure of any cause
      2) Tricuspid regurgitation
         a) Bacterial endocarditis with valvular vegetations
         b) Ebstein's anomaly
      3) Pericardial tamponade, constrictive pericarditis
   c. SVC thrombosis with occlusion below the azygos vein
   d. Azygos continuation of congenital interruption of the inferior vena cava
3. Enlargement of the azygos vein does not occur in patients that have only increased pulmonary blood flow from acyanotic left-to-right shunts. This is another distinguishing feature between overcirculation caused by plasma volume overload and that caused by congenital left-to-right shunts.

D. R4 – enlarged right atrium (RA)
   1. Enlargement of the RA can be determined by assessing the midline-to-right (MR) distance and midline-to-left (ML) distance of the heart.
      a. Draw a vertical line over the midline of the spine (the spinous processes) at the level of the heart
      b. The MR distance is defined as the distance from the midline to the point of maximum convexity of the right heart border. The ML distance is the distance from the midline to the point of maximum convexity of the left heart border.
      c. Normally the MR distance is less than the ML distance. Because the RA is a modified thin-walled venous structure it will distend in response to increased pressure or blood volume. The left ventricle (LV), being a thick-walled muscular structure, will not dilate acutely in response to increased pressure or systemic blood volume.
      d. In the presence of right atrial enlargement the MR distance becomes greater than normal compared to baseline chest radiographs.
   2. Causes of enlargement of the RA
      a. Increased systemic blood volume (plasma volume overload)
         1) Iatrogenic
         2) Renal failure
      b. Increased pressure in the RA
         1) RV failure of any cause
         2) Tricuspid regurgitation
            a) Bacterial endocarditis with valvular vegetations
            b) Ebstein's anomaly
      c. Left-to-right shunts
         1) Atrial septal defect (ASD) — common

        2) Atrioventricular canal (AV canal/endocardial cushion defect) — common

        3) Ventricular septal defect (VSD) — uncommon; occurs only if there is superimposed congestive failure

        4) Patent ductus arteriosus (PDA) — uncommon; occurs only if there is superimposed congestive failure

  d. Admixture lesions

        1) Transposition of the great vessels

        2) Totally anomalous pulmonary venous return — cardiac, supracardiac

  e. Right-to-left shunts at the atrial level

        1) Pulmonic stenosis or atresia with intact ventricular septum

        2) Tricuspid atresia or stenosis, Ebstein's anomaly

E. L1 – enlarged or prominent main pulmonary artery (MPA)

  1. Normal alterations in the MPA with age

    a. In infancy the MPA is covered by the thymus and is obscured.

    b. During childhood the thymus diminishes in prominence and a mild convexity appears at the MPA segment.

    c. In adolescents the MPA segment is convex; this is greatest in females.

    d. In young adults the MPA segment becomes flattened.

    e. After age 35 years the MPA segment becomes concave.

  2. Causes of an enlarged or prominent MPA

    a. Valvular pulmonic stenosis with post-stenotic dilatation related to jet phenomenon

        1) This is associated with enlargement of the left pulmonary artery.

        2) Pulmonary blood flow is usually symmetrical in both lungs; however, because the jet of blood is directed toward the left pulmonary artery, blood flow may be increased in the left lung compared to the right lung.

    b. Increased pulmonary artery pressure

        1) Acute pulmonary arterial hypertension (PAH) such as may occur with massive pulmonary embolism of any cause

        2) Chronic PAH of any cause

    c. Increased pulmonary flow in acyanotic left-to-right shunts

        1) ASD

        2) VSD

        3) PDA

        4) AV canal

    d. Some cyanotic admixture lesions

        1) Truncus arteriosus type I

        2) Totally anomalous pulmonary venous return without venous obstruction

    e. Idiopathic dilatation of the pulmonary artery

        1) This condition is seen in adolescent and young adult females.

        2) Some have mild pulmonic regurgitation, but most patients are asymptomatic.

        3) The left pulmonary artery is not enlarged and blood flow is symmetrical to both lungs.

    f. Congenital absence of the left pericardium

        1) The heart is typically levopositioned.

        2) Either the MPA or the left atrial appendage (LAA) may be prominent because either can protrude through the pericardial defect. Prominence of the MPA is most common.

        3) CT or MR shows lung situated between either the aorta and MPA (most common) or between the MPA and LAA (normally the left pericardium keeps lung out of these recesses).

  3. One caution about the MPA segment on plain films. Any condition that causes the heart to rotate posteriorly to the left will cause a larger portion of the MPA to become visible along the upper left heart border and will cause a prominent convexity in the MPA segment. Conditions which can cause this include: left lower lobe collapse, left lower lobe resection, hypoplasia of the left lung, Swyer-James syndrome, marked pleural thickening on the left side, and congenital or occasionally surgical absence of the left pericardium.

F. L2 – enlarged left pulmonary artery

  1. Valvular pulmonic stenosis

    a. The combination of an enlarged MPA and left pulmonary artery is 99% specific for valvular pulmonic stenosis.

    b. Major differential diagnosis is idiopathic dilatation of the main pulmonary artery.

  2. Congenital interruption of the right pulmonary artery

G. L3 – enlarged left atrium (LA)

  1. The upper limit of normal for the AP dimension of the LA is 4 cm. Although this can be measured on echocardiography and magnetic resonance imaging of the heart, the AP dimension of the left atrium cannot be measured on plain chest radiographs. An oblique LA dimension, however, can be measured on chest radiographs

  2. Method of measuring oblique LA dimension on PA chest radiographs

    a. Draw a diagonal line between the mid-point of the right lateral aspect of the retrocardiac double-density produced by the LA behind the right side of the heart and the midpoint of the left mainstem bronchus

    b. 7 cm or less is normal

    c. Over 7 cm indicates left atrial enlargement

    d. If the LA is massively enlarged it may form the right heart border.

  3. Causes of an enlarged LA

    a. Mitral stenosis or regurgitation

    b. LA myxoma

    c. LV failure

    d. Some acyanotic L-R shunts (but not ASD or AV canal)

        1) VSD

        2) PDA
   e. Cyanotic admixture lesions
        1) Transposition of the great vessels
        2) Truncus arteriosus

H. L4 – enlarged or prominent left atrial appendage (LAA)
   1. Rheumatic mitral stenosis (almost pathognomonic)
   2. LA myxoma (rare)
   3. Supraventricular tachycardia (rare)
   4. Congenital absence of the left pericardium (rare)

I. L5 – left ventricular (LV) stress analysis
   1. The LV stress analysis is a specific method of evaluating the chest radiograph for abnormalities related to the coronary arteries and LV.
   2. The LV stress analysis consists of six separate steps.

J. Six steps of the LV stress analysis
   1. L5 – step 1: aortic valvular calcification
      a. On the lateral view specifically look for calcification in the aortic valve. You will never see it if you don't look for it!
      b. Radiographic distinction between aortic and mitral valvular calcification
        1) On the PA chest radiograph aortic valve calcification lies over the spine and may be hidden, whereas mitral valve calcification usually lies to the left of the spine.
        2) On the lateral chest radiograph draw a line between the end-on left mainstem bronchus and the anterior costophrenic angle. Aortic valve calcification usually lies above this line in the middle third of the heart, whereas mitral valve calcification usually lies below this line in the posterior third of the heart. This method is usually reliable unless the thoracic cage is markedly distorted or there is right-sided chamber enlargement.
      c. Calcification in the aortic valve is specific for calcific aortic valvular stenosis.
        1) Calcification of the valve occurs in about 90% of cases of aortic stenosis.
        2) Calcification is frequently not visible on plain films.
        3) Non-contrast enhanced CT is the best imaging modality for detection of aortic valve calcification.
      d. Aortic stenosis (AS)
        1) Unicommisural valve (1 cusp) – rare
        2) Bicuspid valve (2 cusps)
           a) Most common cause of calcific AS
           b) 1-2% of population
           c) Large posterior (noncoronary) cusp
           d) Right and left cusp fused and ill-defined
           e) Outcome of bicuspid aortic valve
              i) May remain normal
              ii) May develop aortic valvular regurgitation

iii) AS with calcification beginning in middle age

  f) Associated with coarctation of aorta

 3) Tricuspid aortic valve (3 cusps)

  a) Rheumatic aortic stenosis

   i) Rheumatic AS is almost always associated with other diseased valves, typically the mitral valve.

   ii) Isolated AS is almost never rheumatic; rather it indicates either bicuspid aortic valve or degenerative AS.

  b) Degenerative AS — common after age 65

  c) Congenital AS with severe dysplasia of a tricuspid aortic valve — usually presents with severe congestive heart failure shortly after birth

 4) Radiographic signs of AS

  a) Calcification in valve

  b) Enlarged LV

  c) Poststenotic dilatation of ascending aorta

2. L5 – step 2: analysis of LV wall

 a. Hoffman-Rigler sign of LV enlargement

  1) On the lateral chest radiograph identify the point at which the posterior margin of the LV crosses the posterior margin of the inferior vena cava (IVC).

  2) The posterior margin of the LV normally lies 1.8 cm or less from the posterior margin of the IVC at a level 2 cm cephalad to their crossing

  3) If the posterior margin of the LV falls 2 cm or more from the posterior margin of the IVC at a level 2 cm cephalad to their crossing, the LV is likely to be enlarged.

  4) False-positives can occur if the patient is rotated on the lateral view or if there is RV enlargement that displaces the LV posteriorly.

 b. Calcification in LV wall

  1) Calcified myocardial infarcts — the calcification usually lies 2 mm or more from the adjacent lung

  2) Calcified LV aneurysms — due to myocardial thinning the calcification often lies less than 2 mm from the adjacent lung

 c. Abnormal bulges of LV contour

  1) If small and focal, may represent a dyskinetic segment of myocardium.

  2) LV aneurysms tend to be larger and more broad-based against the cardiac silhouette.

  3) LV aneurysms: 2 types

   a) True aneurysm of LV

    i) Contains thinned, dysfunctional LV wall

    ii) Usually involves LV apex

    iii) Anteriorly located on lateral view

    iv) Associated with arrhythmias

        v) Low incidence of rupture

    b) False aneurysm of LV

        i) Represents post-myocardial infarction rupture of LV contained only by pericardium

        ii) Posteriorly located on lateral view

        iii) High incidence of rupture and sudden death

3. **L5 – step 3: coronary artery triangle**

    a. The coronary artery triangle is the space on the PA chest radiograph defined by the lateral margin of the spine medially, the left pulmonary artery superiorly, and the midpoint of the left side of the heart inferiorly.

    b. The coronary artery triangle contains the left main coronary artery, left anterior descending coronary artery, and the circumflex coronary artery. Search this area closely for calcification.

    c. Since the coronary arteries arise from the aortic root, coronary artery calcification also may be seen on the lateral view in the region near the aortic valve.

    d. Coronary artery calcification appears as parallel tram-track calcification or as circular calcification if seen end-on.

    e. Radiographic detection of coronary artery calcification

        1) Plain films have low sensitivity but high specificity. Plain film visualization of coronary artery calcification indicates significant coronary artery disease.

        2) Fluoroscopy has better sensitivity and can be done when plain film findings are equivocal or suspicious.

        3) CT, particularly fast-CT (electron beam CT, multi-slice helical CT), is the best imaging method for detecting coronary artery calcification.

4. **L5 – step 4: enlarged ascending aorta**

    a. Hypertension

    b. Atherosclerotic ectasia

    c. Post-stenotic dilatation in AS

    d. Cystic medial necrosis of the ascending aorta (annuloaortic ectasia)

    e. Coarctation of the aorta

    f. Dissecting hematomas

    g. True aneurysms (luetic, mycotic, and atherosclerotic)

    h. Traumatic pseudoaneurysms

5. **L5 – step 5: abnormal aortic knob**

    a. Hypertension

    b. Atherosclerotic ectasia

    c. Coarctation of the aorta

        1) Two main forms

            a) Localized (postductal or "adult" type)

                i) Most common form

                ii) Usually presents during childhood, adolescence, or later life

            b) Tubular hypoplasia (diffuse, preductal, or "infantile" type)

     i) Long segment of hypoplasia of the aortic arch after the origin of the innominate artery

     ii) Usually presents with congestive heart failure after the ductus closes; typically in the second or third week of life

  2) Approximately 50% of patients with coarctation have associated bicuspid aortic valve (note: published figures range from 25-85% but the 50% figure is the most commonly quoted).

  3) Also associated with Turner's syndrome — about 45% of children with Turner's syndrome have coarctation of the aorta

  4) Severe coarctation may clot and resemble congenital interruption of the aorta.

  5) Primary signs of coarctation

    a) Notch or "3" sign

    b) Reverse "3" or "E" sign on barium swallow

    c) The aortic knob may be either small or enlarged

    d) The left subclavian artery is almost always enlarged

    e) Post-stenotic dilatation of the descending aorta

  6) "Collateral" signs of coarctation

    a) Superior mediastinal widening from collateral arteries

    b) Rib notching

     i) Rib notching is rarely visible before age 8 years. After age 8 the incidence of rib notching increases with increasing age. About 75% of adults with untreated coarctation of the aorta will have rib notching.

     ii) The first two ribs are usually spared. The first and second intercostal arteries originate from the superior intercostal artery, which arises from the subclavian artery above the coarcted segment, and do not participate in the collateral flow. The tenth, eleventh, and twelfth ribs are also spared because their intercostal arteries do not communicate with the internal mammary arteries anteriorly.

     iii) Rib notching involves the third through the ninth ribs. The most prominent rib notching involves the fourth through eighth ribs.

     iv) Involves inferior aspect of ribs, which may also be sclerotic

    c) Dilated internal mammary arteries may be seen on lateral chest radiograph as lobular retrosternal opacities.

    d) Angiography or magnetic resonance imaging are useful in documenting the coarcted segment and collateral vessels.

 d. Pseudocoarctation of the aorta

  1) Marked kinking of the aorta near isthmus region

  2) Pseudocoarctation was once thought to be secondary to atherosclerosis; however, several cases have been reported in children.

  3) Pseudocoarctation most likely represents a mild form of coarctation that is not hemodynamically significant; therefore there

are no collateral vessels. This is the distinguishing feature between coarctation and pseudocoarctation.

    4) Typically the luminal constriction is less than 50% and the pressure gradient across the constricted area is less than 10 mm Hg.

    5) May closely resemble coarctation on plain films

  e. Dissecting hematomas

  f. True aneurysms (mycotic, atherosclerotic)

  g. Traumatic pseudoaneurysms

6. **L5 – step 6: enlarged descending aorta**

  a. Hypertension

  b. Atherosclerotic ectasia

  c. Aortic valvular regurgitation

  d. Dissecting hematomas

  e. True aneurysms (mycotic, atherosclerotic)

  f. Traumatic pseudoaneurysms

References and suggested additional reading

1. Baron MG. The cardiac silhouette. J Thorac Imaging 2000; 15:230-242.
2. Dähnert W. *Radiology Review Manual 4th ed.* Philadelphia: Lippincott Williams & Wilkins, 2000.
3. Elliott LP. Cardiovascular anatomy stage of analysis. In: Taveras JM, Ferrucci JT eds. *Radiology: Diagnosis-Imaging-Intervention.* Volume 2, Chapter 12. Hagerstown: Lippincott, 1991.
4. Elliott LP. The plain film approach utilizing differential diagnostic vascular anatomical sets. In: Taveras JM, Ferrucci JT eds. *Radiology: Diagnosis-Imaging-Intervention.* Volume 2, Chapter 13. Hagerstown: Lippincott, 1991.
5. Elliott LP. Introduction to diseases showing normal vascularity and a normal size heart. In: Taveras JM, Ferrucci JT eds. *Radiology: Diagnosis-Imaging-Intervention.* Volume 2, Chapter 14. Hagerstown: Lippincott, 1991.
6. Fuster V, Alexander RW, O'Rourke RA, Roberts R, King SB III, Wellens HJJ. *Hurst's The Heart 10th ed.* New York: McGraw-Hill, 2001.
7. Gedgaudas E, Moller JH, Castaneda-Zuniga WR, Amplatz K. *Cardiovascular Radiology.* Philadelphia: Saunders, 1985.
8. Higgins CB, Reinke RT, Jones NE, Broderick T. Left atrial dimension on the frontal thoracic radiograph: a method for assessing left atrial enlargement. AJR 1978; 130:251-255.
9. Hoffman RB, Rigler LG. Evaluation of left ventricular enlargement in the lateral projection of the chest. Radiology 1965; 85:93-100.
10. Keats TE, Sistrom C. *Atlas of Radiologic Measurement, 7th ed.* St. Louis: Mosby, 2001.
11. Meszaros WT. *Cardiac Roentgenology: Plain Films and Angiocardiographic Findings.* Springfield: Charles C Thomas, 1969.
12. Miller SW. *Cardiac Radiology: The Requisites.* St. Louis: Mosby, 1996.

John H. Woodring, M.D.

13. Spindola-Franco H, Fish BG. *Radiology of the Heart: Cardiac Imaging in Infants, Children, and Adults*. New York: Springer-Verlag, 1985.
14. Steiner RE. Radiology of the pulmonary circulation. In: Abrams HL ed. *Abrams Angiography Vascular and Interventional Radiology 3rd ed*. Boston: Little, Brown, 1983:763-781.
15. Woodring JH, Rhodes RA III. Posterosuperior mediastinal widening in aortic coarctation. AJR 1985; 144:23-25.
16. Woodring JH, West JW. CT of aortic and mitral valve calcification. J Ky Med Assoc 1989; 87:177-180.

Chapter 2.

Cardiac conditions presenting with a normal-sized heart and normal pulmonary vascularity

I. **Determination of normal pulmonary vascularity in adults**
   A. **Blood flow should be symmetrical and equal in both lungs.**
   B. **Examine the right hilar angle**
      1. The right hilar angle is formed by the right superior pulmonary vein as it crosses in front of the descending branch of the right pulmonary artery to enter the left atrium.
      2. The right hilar angle is normally acute.
      3. An acute left hilar angle is occasionally formed by the left superior pulmonary vein as it crosses the descending branch of the left pulmonary artery. However, because the cardiac silhouette often partially obscures the vascular structures in the left hilum and medial aspect of the left lower lobe, determination of the pulmonary vascular pattern is primarily based on the right lung.
   C. **The descending branch of the right pulmonary artery should be <2 cm in diameter.**
      1. The transverse width of the descending branch of the right pulmonary artery is usually measured just below the right hilar angle.
      2. The upper limit of normal for the width of the descending branch of the right pulmonary artery is 1.5 cm in women and 1.6 cm in men.
      3. If the descending branch of the right pulmonary artery measures 2 cm or more in width the descending branch of the right pulmonary artery is abnormally enlarged.
   D. **Examine the distribution of blood flow in the lungs**
      1. Divide the lungs into an upper lung zone and a lower lung zone with the right hilar angle being the dividing line between the two zones.
      2. Because of gravity the vessels in the lower lung zones should be larger than those in the upper lung zones.
         a. In the upright position a hydrostatic pressure gradient is established in the pulmonary arterial and venous circulations with hydrostatic pressure being greatest in the lung base and progressively diminishing as one moves toward the lung apex. The average height of the adult lung is about 30 cm; therefore, the difference in hydrostatic pressure between the apex and base is approximately 30 cm H_2O or about 20 mm Hg.
         b. The hydrostatic pressure gradient in the pulmonary arteries in the upright position ranges from approximately 25 mm Hg at the lung base to 5 mm Hg at the lung apex with a mean value of about 15 mm Hg.

c. The hydrostatic pressure gradient in the pulmonary veins in the upright position ranges from 15 mm Hg at the base to –5 mm Hg at the lung apex with a mean value of about 5 mm Hg.

d. Although the pressures vary greatly between the lung apex and the base because of the weight of the blood itself, the difference in pressure between the arteries and veins, the driving pressure, is constant at all levels since the hydrostatic increments are equal in the two vascular systems. The driving pressure in the pulmonary circulation is 10 mm Hg.

e. Since intraalveolar pressure can be assumed to be constant over the 30 cm height of the lung, pulmonary blood flow is determined almost exclusively by the hydrostatic pressure gradient in the lung.

   1) Pulmonary blood flow is greatest in the lung base because pulmonary arterial and venous pressures are both greater than intraalveolar pressure and blood flow is free and unrestricted.

   2) As one moves from the lung base toward the lung apex pulmonary arterial and venous pressures both drop. In the upper lung zone intraalveolar pressure exceeds venous pressure and the pulmonary veins are collapsed. However, pulmonary arterial pressure still exceeds intraalveolar pressure in the upper lung zone and there is enough pressure to perfuse the upper lobe vessels during right ventricular systole. Pulmonary venous return from the upper lung zone drops down to the left atrium in a "waterfall" fashion.

   3) Since arterial pressure progressively decreases as one moves from the lung base to the apex, pulmonary perfusion and blood flow also progressively decrease as one moves from the lung base to the apex.

f. The pulmonary arteries are thin-walled, contain little smooth muscle, and are easily distensible. These characteristics of the pulmonary arteries result in a remarkably low-pressure system, with low resistance to flow, that can accommodate the entire right ventricular output (5 L/min) at a much lower pressure than that in the systemic circulation. Because of the easy distensibility of the pulmonary circulation there is a close relationship between pulmonary blood flow and vessel diameter.

   1) The flow-caliber relationship can be described by the formula, flow (cc/min) = D^2 X 10, where D is the vessel diameter.

      a) For example, a subsegmental pulmonary artery 1.5 mm in diameter would have a blood flow of 22.5 cc/min.

      b) A pulmonary artery 5 mm in diameter would have a flow of 250 cc/min.

   2) Because of this flow-caliber relationship in the pulmonary circulation, regional differences in hydrostatic pressure and pulmonary blood flow are reflected on the upright chest radiograph by regional differences in the size of the visible pulmonary arteries

and veins, with the pulmonary vessels being larger in the lower lung zone than those in the upper lung zone. This results in a normal blood flow ratio (upper zone vessel size divided by lower zone vessel size) that is <1.

3. It is of considerable importance to know specifically which pulmonary vessels should be examined on the frontal chest radiograph in order to determine whether or not the pulmonary blood flow pattern is normal.

   a. One must compare vessels in the upper lung zone and lower lung zone that are of a similar order of branching from the central right pulmonary artery.

   b. The upper zone vessels used in this comparison are the segmental and subsegmental arteries and veins in the right upper lobe immediately above the right upper lobe bronchus. These vessels are encompassed by a small circle about 4-5 cm in diameter drawn on the frontal chest radiograph bounded inferiorly by the right upper lobe bronchus and medially by the superior vena cava.

   c. The lower zone vessels used in this comparison are the segmental and subsegmental arteries and veins in the right lower lobe immediately below the descending branch of the right pulmonary artery. These vessels are encompassed by a small circle about 4-5 cm in diameter drawn on the frontal chest radiograph in the right cardiophrenic angle bounded inferiorly by the apex of the right hemidiaphragm and medially by the right atrium.

   d. If the lower zone vessels are visibly larger than the upper zone vessels on an upright posteroanterior chest radiograph, pulmonary blood flow is normal.

   e. The lateral chest radiograph can also be used to evaluate the vascular pattern. In a normal subject the pulmonary vessels in the lower lobes posterior to the heart are noticeably larger than the pulmonary vessels in the upper lobes above the hila on the lateral view.

4. The pulmonary artery-bronchus ratio (ABR) may be a helpful adjunct in determining the pulmonary vascular pattern.

   a. The paired pulmonary arteries and bronchi run in a shared connective tissue sheath. In the absence of a gravitational hydrostatic pressure gradient (for example, in a supine subject) the paired end-on pulmonary arteries and end-on bronchi are roughly the same size. The ABR, defined as the external diameter of an end-on pulmonary artery divided by the external diameter of its accompanying bronchus, would be approximately 1.

   b. In the upright position, regional differences in pulmonary blood flow occur, but the bronchi are unchanged in size. This causes the lower lung zone pulmonary arteries to become larger than their accompanying bronchi (ABR >1) and the upper lung zone pulmonary arteries to become smaller than their accompanying bronchi (ABR <1).

5. Normal pulmonary vascularity on supine radiographs.
    a. For all intents and purposes the discussion of normal pulmonary vascularity is limited to the discussion of upright posteroanterior (PA) and lateral chest radiographs.
    b. In the supine position the hydrostatic pressure gradient in the pulmonary circulation is eliminated. Furthermore, blood volume in the lungs is increased by approximately 30% due to blood draining upward from the lower extremities. This causes a reorientation of pulmonary blood flow with flow being equal in the upper and lower lung zones with a blood flow ratio of 1.
    c. The end-on pulmonary arteries and their accompanying end-on bronchi are roughly of the same size in both the upper and lower lung zones with an ABR of 1 in both the upper and lower lung zones.

II. Determination of normal pulmonary vascularity in infants
A. Blood flow should be equal in both lungs.
B. The transverse width of the descending branch of the right pulmonary artery should be equal to, or less than, the transverse internal diameter of the trachea.
1. The descending branch of the right pulmonary artery is measured below the right hilar angle. The transverse internal diameter of the trachea is measured above the impression of the aorta.
2. This sign may not be useful in some newborns because the thymus may obscure the right hilum.
3. Is often the best method for determining normal vascularity in older infants and children
C. Evaluate the pulmonary vessels that can be seen through the liver.
1. Branching pulmonary vessels seen through the liver, below the apex of right hemidiaphragm, should be barely visible.
2. End-on vessels seen through the liver should be minimally perceptible or not visible at all.
3. This method for determining normal pulmonary vascularity is particularly helpful in newborns and small infants in which the thymus or cardiac silhouette obscures the descending branch of the right pulmonary artery.

III. Determination of normal cardiac size
A. The cardiothoracic ratio (CTR)
1. Typically measured on erect PA chest radiographs
2. The CTR is defined as the maximum transverse diameter of the heart divided by the maximum internal diameter of the thorax at or near the apex of the right hemidiaphragm.
    a. The maximum transverse diameter of the heart is determined by adding the midline-to-right (MR) distance and midline-to-left (ML) distance of the heart.
        1) Draw a vertical line over the midline of the spine (the spinous processes) at the level of the heart.

16

2) The MR distance is defined as the distance from the midline to the point of maximum convexity of the right heart border.

3) The ML distance is the distance from the midline to the point of maximum convexity of the left heart border.

b. The maximum internal diameter of the thorax is the widest portion of the lower thorax. Although this point is often at or near the level of the apex of the right hemidiaphragm, the measurement should be made at the widest point of the lower thorax, not necessarily at the exact level of the apex of the right hemidiaphragm.

3. Variations of CTR with age:
   a. 0-1 yr.   0.65
   b. 1-2 yrs.   0.60
   c. 2-4 yrs.   0.52
   d. 5 and over   0.50

B. CTR on supine anteroposterior (AP) films

1. On a supine AP chest radiograph the cardiac silhouette is magnified more than the rib cage.
2. This causes the heart to visually appear quite large, especially on a film obtained with a short focal-film-distance.
3. There is approximately a 15% increase in the measured CTR because of the relatively greater magnification of the heart compared to the rib cage.
4. Therefore, the upper limit of normal for the CTR on a supine AP chest films is 58%.

IV. Differential diagnosis of normal-sized heart with normal vascularity — mnemonic = "MANO"

A. Myocardial ischemia

1. Specifically search the coronary artery triangle for evidence of coronary artery calcification.
   a. Parallel tram-track calcification
   b. Circular calcification seen end-on
2. Since the coronary arteries arise from the aortic root, coronary artery calcification may also be seen on the lateral view near the aortic valve location (middle third of cardiac silhouette above a line drawn between the end-on left mainstem bronchus and the anterior cardiophrenic angle).
3. Carefully search the left ventricular (LV) wall for evidence of calcified myocardial infarcts or LV aneurysms.

B. Afterload stress

1. Conditions placing afterload stress on the right side of the heart.
   a. Valvular pulmonic stenosis
      1) 9% of cases of congenital heart disease
      2) Congenital dome-like narrowing of the pulmonic valve
      3) Enlarged main pulmonary artery
      4) Enlarged left pulmonary artery
      5) Right ventricular hypertrophy

6) Some have increased vascularity on the left but most have normal flow bilaterally.

7) Associated with Noonan's syndrome (male Turner syndrome, pseudo-Turner syndrome, male Ullrich syndrome, Turner phenotype, or XX or XY Turner phenotype)
   a) Short stature
   b) Mental retardation
   c) Delayed puberty
   d) Hypertelorism
   e) Low-set ears
   f) Triangular face
   g) Ptosis
   h) Webbed neck
   i) Dental malocclusion
   j) Pectus carinatum with distal pectus excavatum
   k) Undescended testes
   l) Valvular pulmonic stenosis
   m) Ventricular septal defect
   n) Other cardiac abnormalities including supravalvular pulmonic stenosis, peripheral pulmonic stenosis, atrial septal defect, dysplasia of the aortic valve, and hypertrophic cardiomyopathy
   o) May have familial transmission
   p) Normal chromosomal analysis

b. Supravalvular/peripheral pulmonic stenosis
   1) This condition is characterized by one or more areas of stenosis of the pulmonary arteries distal to the pulmonic valve.
   2) The chest radiograph may be normal or may show signs of pulmonary arterial hypertension. In some cases the pulmonary arteries have a "beaded" or "sausage-like" appearance because of areas of stenosis with poststenotic dilatation of the arteries distal to the areas of stenosis.
   3) The diagnosis is usually made by pulmonary arteriography.

2. Conditions placing afterload stress on the left side of the heart.
   a. Hypertension
      1) The chest radiograph is usually normal.
      2) If hypertension is long-standing and severe, there will be left ventricular enlargement (LVE) and enlargement of the aorta.
         a) These findings are nonspecific and can also be seen with atherosclerotic cardiovascular disease and aortic valve disease.
         b) In the past the combination of LVE and enlargement of the aorta was referred to as the "hypertensive-aortic configuration" although this term neglected atherosclerotic cardiovascular disease, which is very common.
         c) If you have LVE and enlargement of the aorta the differential diagnosis includes hypertension, atherosclerotic cardiovascular

disease, and aortic valve disease. If you can find coronary artery calcification or aortic valve calcification a specific diagnosis can be offered, but if no characteristic calcifications are present, you are left with a relatively short differential diagnostic list.

b. Aortic stenosis — look for characteristic valvular calcification and prominence of the ascending aorta

c. Coarctation of the aorta — look for characteristic primary or collateral signs of coarctation

C. <u>N</u>ormal

1. "In the absence of roentgen findings, no diagnosis is possible." (Larry Elliott, MD, 1986)

2. Many chest radiographs are normal, and the chest radiographs of many patients with cardiac disease show no abnormality. The purpose of this mnemonic is to force yourself to look closely at what may at first appear to be a normal chest radiograph to make sure that you are not missing specific radiographic signs that might permit a specific cardiac diagnosis to be made. If no signs of cardiac disease are present, then the chest radiograph would be interpreted as normal or negative.

3. Normal films are quite common in clinical practice, but normal films are seldom shown on oral board exams. If you think you are being shown a normal radiograph on the oral board exam you are probably missing something — go back and systematically search the film again.

D. <u>O</u>ther (miscellaneous)

1. Congenital absence of the pericardium

2. Mitral annulus calcification
   a. Common after age 65
   b. Occurs more often in women
   c. Represents degenerative calcification of the mitral annulus
   d. Usually seen as "C," "U," or "J" shaped calcification at the location of the mitral valve
   e. Usually clinically insignificant
   f. 15% have mitral regurgitation, mitral stenosis, or mitral valve prolapse — mitral regurgitation is most common

References and suggested additional reading

1. Best CH, Taylor NB. *The Physiological Basis of Medical Practice, 8th ed.* Baltimore: Williams and Wilkins, 1966.

2. Coussement AM, Gooding CA. Objective radiographic assessment of pulmonary vascularity in children. Radiology 1973; 109:649-654.

3. Elliott LP. Introduction to diseases showing normal vascularity and a normal size heart. In: Taveras JM, Ferrucci JT eds. *Radiology: Diagnosis-Imaging-Intervention.* Volume 2, Chapter 14. Hagerstown: Lippincott, 1991.

4. Keats TE, Sistrom C. *Atlas of Radiologic Measurement, 7th ed*. St. Louis: Mosby, 2001.
5. Meszaros WT. *Cardiac Roentgenology: Plain Films and Angiocardiographic Findings*. Springfield: Charles C Thomas, 1969.
6. Miller SW. *Cardiac Radiology: The Requisites*. St. Louis: Mosby, 1996.
7. Milne ENC, Pistolesi M. *Reading the Chest Radiograph: a Physiologic Approach*. St. Louis: Mosby, 1993.
8. Müller NL, Fraser RS, Colman NC, Paré PD. *Radiologic Diagnosis of Diseases of the Chest*. Philadelphia: Saunders, 2001.
9. Simon M. Physiologic considerations in radiology of the pulmonary vasculature. In: Abrams HL ed. *Abrams Angiography Vascular and Interventional Radiology 3rd ed*. Boston: Little, Brown, 1983:783-802.
10. Spindola-Franco H, Fish BG. *Radiology of the Heart: Cardiac Imaging in Infants, Children, and Adults*. New York: Springer-Verlag, 1985.
11. Steiner RE. Radiology of the pulmonary circulation. In: Abrams HL ed. *Abrams Angiography Vascular and Interventional Radiology 3rd ed*. Boston: Little, Brown, 1983:763-781.
12. West JB. *Respiratory physiology—the Essentials*. Baltimore: Williams and Wilkins, 1974.
13. Woodring JH. Pulmonary artery-bronchus ratios in patients with normal lungs, pulmonary vascular plethora, and congestive heart failure. Radiology 1991; 179:115-122.
14. Woodring JH, Phillips BA, West JW, Ulmer J, Cooper JK. A prospective evaluation of plain radiographic signs of chronic obstructive pulmonary disease. J Thorac Imaging 1991; 6:14-21.

Chapter 3.

Cardiac conditions presenting with an enlarged heart and normal pulmonary vascularity

I. **Differential diagnosis of an enlarged heart with normal vascularity —
mnemonic = "2PICA" or "PPICA"**
   A. **Pericardial diseases**
      1. The normal pericardium
         a. The normal pericardium is usually not visible on upright posteroanterior views; however, the normal pericardium is often visible in front of the heart on the lateral view.
            1) The apposed visceral and parietal pericardium are seen on the lateral view as a vertical white line in front of the heart outlined by the epicardial fat on the surface of the heart posteriorly and by the pericardial fat pad anteriorly.
            2) This vertical white line formed by the pericardium anterior to the heart is called the anterior pericardial stripe.
            3) The anterior pericardial stripe is normally 1-2 mm in thickness.
         b. The normal pericardium is usually well-visualized on CT and MR and again is normally 1-2 mm in thickness.
      2. Pericardial effusion
         a. Causes of pericardial effusion
            1) Noninfectious causes
               a) Acute idiopathic
               b) Acute myocardial infarction
               c) Left ventricular (LV) failure
               d) Uremia
               e) Hypoalbuminemia
               f) Neoplasia
               g) Trauma
               h) Radiation
               i) Myxedema
               j) Ascending aortic dissection
               k) Anticoagulation
               l) Sarcoidosis
            2) Causes related to hypersensitivity or autoimmunity
               a) Rheumatic fever
               b) Collagen vascular disease
                  i) Rheumatoid arthritis
                  ii) Systemic lupus erythematosus
                  iii) Scleroderma
               c) Drug reaction

          i) Procainamide

          ii) Hydralazine

      d) Postmyocardial infarction (Dressler's syndrome)

      e) Postpericardiotomy

    3) Infectious causes

      a) Bacterial pericarditis

      b) Viral pericarditis

      c) Tuberculosis

      d) Fungal infection

  b. Plain film signs of pericardial effusion

    1) Relatively specific signs of pericardial effusion

      a) "Displaced" epicardial fat pad sign

        i) This sign is really a misnomer because the epicardial fat pad is not displaced — it remains on the epicardial surface of the heart.

        ii) As pericardial fluid accumulates in the pericardial sac, the anterior pericardial stripe becomes thickened and the distance between the epicardial fat pad and the pericardial fat increases causing the epicardial fat pad to appear "inwardly displaced" within the cardiac silhouette on the lateral view.

        iii) If the anterior pericardial stripe is 4 mm or greater in thickness it is abnormal. This is usually due to pericardial effusion but can also be seen in pericardial thickening.

        iv) Occurs in 50-57% of cases

      b) Posteroinferior bulge sign

        i) Early pericardial effusion typically accumulates in the dependent portion of the pericardial sac along the diaphragmatic surface and in the posteroinferior pericardial recess.

        ii) On the lateral view pericardial fluid accumulating in the posteroinferior pericardial recess causes a characteristic alteration in the posteroinferior aspect of the cardiac silhouette — the posteroinferior margin of the cardiac silhouette bulges posteriorly and inferiorly toward the spine forming an obtuse angle with the left hemidiaphragm.

        iii) Occurs in up to 87% of cases

        iv) Can be distinguished from LV enlargement (LVE) by the fact that LVE usually does not occur acutely and an enlarged LV usually forms an acute angle with the left hemidiaphragm

      c) Water-bottle configuration

        i) The cardiac silhouette is enlarged and broadened at the diaphragmatic level often forming a very acute angle with the right hemidiaphragm.

        ii) This is similar to the shape assumed by a leather water-bottle filled with water when it is placed on a table.

         iii) Occurs only in moderate-to-large pericardial effusions

         iv) Seen in about 40% of cases

    d) Widening of the carinal angle

         i) Widening of the carinal angle can be seen in pericardial effusion, left atrial enlargement (LAE), and with subcarinal masses.

         ii) In a patient with an enlarged cardiac silhouette and no specific signs of LAE you should consider pericardial effusion as a cause and search for other signs.

         iii) Occurs in about 18% of cases

    e) Differential-density sign

         i) In large pericardial effusions there may be enough fluid present to actually demonstrate a difference in density between the soft tissue density of the heart and the lesser density of the pericardial fluid surrounding it.

         ii) Occurs in about 13% of cases

2) Nonspecific signs of pericardial effusion

    a) Enlargement of the cardiac silhouette

         i) Any acute enlargement of the cardiac silhouette should suggest the possibility of pericardial effusion.

         ii) Occurs in about 90% of cases

    b) Hyperacute right cardiophrenic angle (Rotch sign) — is useful when it is part of the water-bottle configuration; however, a hyperacute right cardiophrenic angle by itself is of little diagnostic value

    c) Loss of retrosternal clear space — occurs in large pericardial effusions, but also occurs in right-sided chamber enlargement, dilatation of the ascending aorta, dilatation of the main pulmonary artery, and also in some normals

    d) Straight left heart border

         i) Frequently mentioned in literature

         ii) Occurs in early pericardial effusions, early LAE, and in many normal subjects

         iii) In moderate-to-large pericardial effusions the left heart border is often convex.

         iv) Is of little or no diagnostic value

c. Pulmonary vascular patterns and pericardial effusion

1) Pericardial effusion is presented here, in the section on an enlarged heart with normal pulmonary vascularity, because in the majority of cases that is how pericardial effusion presents radiographically.

2) If the cause of pericardial effusion is LV failure, one might expect to see a vascular pattern of pulmonary venous hypertension.

3) If the cause of pericardial effusion is uremic pericarditis one might expect to see a vascular pattern of overcirculation of the pulmonary vascularity.

      4) If there is pericardial tamponade the pulmonary vascularity may be decreased.

  d. CT and MR signs of pericardial effusion

      1) On CT pericardial effusion appears as homogeneous water-attenuation material surrounding the heart.

      2) On spin echo MR images simple transudative pericardial effusion has low signal intensity on T1-weighted images and high signal intensity on T2-weighted images. Slow motion of pericardial fluid may occasionally cause medium signal intensity on T1-weighted images. Hemorrhagic or exudative effusions typically have areas of medium or high signal intensity on T1-weighted images. Pericardial effusion has high signal intensity on cine gradient-echo sequences because of motion of the fluid in the pericardial sac.

      3) MR is superior to CT in distinguishing pericardial thickening from pericardial effusion, and in demonstrating fibrinous and hemorrhagic pericardial effusion.

3. Pericardial tamponade

  a. Increased pressure in the pericardial sac causes compression and flattening of the atria resulting in diminished atrial filling.

  b. The pulmonary vascularity is usually normal; however, decreased cardiac filling may result in decreased pulmonary vascularity.

  c. Compression of the right atrium results in distention of azygos vein and superior vena cava (systemic venous hypertension).

4. Constrictive pericarditis

  a. Usually due to inflammatory or neoplastic thickening of the pericardium — tuberculosis is no longer the most common cause

  b. Calcification occurs in 50% and is a finding strongly suggestive of constrictive pericarditis in a patient with the appropriate clinical history; however, pericardial calcification can be present without constriction.

  c. The pericardium is usually 4mm or greater in thickness.

      1) Thickening of the pericardium can be diffuse but is often focal.

      2) Absence of pericardial thickening argues against a diagnosis of constriction but does not completely exclude it.

  d. On plain films the cardiac silhouette is usually enlarged, not normal or small as is often taught.

  e. On CT or MR the ventricles are compressed.

      1) Diffuse thickening often compresses both ventricles.

      2) Focal pericardial thickening may compress only the right ventricle (RV).

  f. There are usually signs of systemic venous hypertension including dilatation of the neck veins, dilatation of the azygos vein and superior vena cava, dilatation of the atria and coronary sinus, dilatation of the inferior vena cava and hepatic veins, passive congestion of the liver, ascites, and pleural effusion.

  g. Since CT is better at showing pericardial calcification than MR, it is the preferred method for evaluating constrictive pericarditis.

    h. The pulmonary vascular pattern is usually normal, but when there is significant compression of the LV there may be pulmonary venous hypertension.

B. Preload stress
1. Conditions that produce preload stress on the LV
    a. Aortic valvular regurgitation
        1) Causes
            a) Rheumatic heart disease
            b) Bicuspid aortic valve
            c) Bacterial endocarditis
            d) Syphilis
            e) Aortitis
                i) Ankylosing spondylitis
                ii) Reiter's syndrome
                iii) Psoriasis
                iv) Systemic lupus erythematosus
                v) Giant cell aortitis
                vi) Takayasu's arteritis
            f) Trauma
            g) Systemic hypertension
            h) Atherosclerotic cardiovascular disease
            i) Aortic dissection
            j) Cystic medial necrosis of the ascending aorta (annuloaortic ectasia) (Marfan's syndrome)
            k) Homocystinuria
            l) Ehlers-Danlos syndrome
            m) Osteogenesis imperfecta
            n) Ventricular septal defect with prolapsing aortic cusp
        2) LV enlargement is a hallmark of aortic regurgitation.
        3) The aorta, and in particular the descending aorta, is often dilated.
            a) This is due to the widened pulse pressure characteristic of aortic regurgitation.
            b) During diastole the LV receives blood from the left atrium (LA) plus that regurgitated from the aorta. Because of this LV stroke volume is greatly increased.
            c) During the first half of systole there is an explosive ejection of a large volume of blood into the aorta resulting in greatly increased systolic pressure. Systolic pressure drops rapidly during the second half of systole. Diastolic pressure is lower than normal in part due to the aortic regurgitation and in part due to peripheral vasodilatation that accompanies aortic regurgitation. The increased volume and systolic pressure in the first half of systole cause aortic dilatation.
            d) The ascending aorta is relatively protected from dilatation compared to the descending aorta because of the aortic

regurgitation and rapid run-off of blood into the brachiocephalic vessels.

4) These findings, however, are nonspecific and can be seen in hypertension and atherosclerotic cardiovascular disease.

5) When aortic regurgitation is secondary to dilatation of the aortic valve annulus, such as occurs in annuloaortic ectasia, only the ascending aorta may be dilated.

6) Aortic regurgitation is usually associated with some degree of aortic stenosis (remember that most stenotic valves do not close completely and are relatively regurgitant, and that most regurgitant valves are relative stenotic for the increased stroke volume flowing through them).

b. Mitral valvular regurgitation

1) Causes
   a) Rheumatic heart disease
   b) Rupture or dysfunction of papillary muscles from myocardial infarction
   c) Rupture of chordae tendineae from bacterial endocarditis or trauma
   d) Prolapse of mitral valve
      i) Myxomatous degeneration of mitral valve
      ii) Ischemia
   e) Hypertrophic cardiomyopathy
   f) Any cause of dilatation of the LV and mitral annulus
      i) Hypertension
      ii) Aortic regurgitation
      iii) Dilated cardiomyopathy
      iv) Atherosclerotic cardiovascular disease

2) The LA and LV are both typically dilated.

3) When the cause is rheumatic heart disease, the chest radiograph frequently shows enlargement of LA appendage from associated mitral stenosis.

4) A mild degree of mitral regurgitation also occurs in a small percentage of patients with degenerative calcification of the mitral annulus.

5) May be associated with pulmonary venous varix
   a) In some cases of long-standing mitral regurgitation one of the pulmonary veins may become markedly enlarged and varicose.
   b) Is thought to be due to a jet of regurgitant blood from the mitral valve that selectively dilates one of the pulmonary veins
   c) Typically involves one of the inferior pulmonary veins
   d) Chest radiographs typically show a lobular density behind the heart closely related to the LA.
   e) The diagnosis can be easily confirmed by CT, which shows the lobular enlarged pulmonary vein entering the LA.

      f) Can be distinguished from arteriovenous malformation or fistula by the fact that it is not attached in any way to the pulmonary artery

C. Ischemia

1. Coronary artery disease — look specifically for signs of ischemic heart disease including coronary artery calcification, calcified myocardial infarcts, and LV aneurysms.
2. Anomalous left coronary artery
   a. The left coronary artery arises from the main pulmonary artery rather than from the aorta.
   b. During fetal life, because of elevated pulmonary artery pressures, the left coronary artery is adequately perfused from the pulmonary artery.
   c. After birth, pulmonary artery pressures fall.
      1) Perfusion of the left coronary artery ceases from the main pulmonary artery.
      2) Perfusion of the left coronary artery then depends on collateral vessels from the right coronary artery, which are scant.
      3) This results in ischemia of the LV.
   d. Infants experience angina with crying and diaphoresis
   e. The LV and subsequently the LA become dilated resulting in a markedly enlarged cardiac silhouette.

D. Cardiomyopathy

1. Dilated cardiomyopathy
   a. Idiopathic
   b. Coronary artery disease
   c. Peripartum cardiomyopathy
   d. Alcohol
   e. Infectious
      1) Viral (coxsackievirus, cytomegalovirus, human immunodeficiency virus)
      2) Tuberculosis
      3) Fungal
      4) Diphtheria
      5) Rickettsia
      6) Parasitic
   f. Drugs
      1) Cocaine
      2) Phenothiazines
      3) Antibiotics
      4) Diuretics
      5) Anticonvulsants
      6) Anti-inflammatory drugs
      7) Doxorubicin, bleomycin, antiretroviral agents
   g. Toxic metals (cobalt, mercury, lead)
   h. Carbon monoxide
   i. Connective tissue diseases

        1) Scleroderma
        2) Dermatomyositis
        3) Systemic lupus erythematosus
    j. Endocrine disorders
        1) Hypothyroidism
        2) Thyrotoxicosis
        3) Cushing's disease
        4) Acromegaly
        5) Diabetes mellitus
        6) Pheochromocytoma
    k. Neuromuscular diseases
        1) Duchenne's muscular dystrophy
        2) Facioscapulohumeral dystrophy
        3) Erb limb-girdle dystrophy
        4) Myotonic dystrophy
        5) Friedreich's ataxia

2. Hypertrophic cardiomyopathy
    a. Genetically transmitted — autosomal dominant with variable penetrance
        1) Obstructive
            a) Subaortic stenosis (formerly idiopathic hypertrophic subaortic stenosis — IHSS)
        2) Nonobstructive
            a) Symmetric hypertrophy
            b) Asymmetric hypertrophy
                i) Asymmetric septal hypertrophy (ASH)
                ii) Apical hypertrophy
    b. Acquired
        1) Hypertension
        2) Renal parenchymal and renovascular disease
        3) Adrenal diseases
        4) Endocrine disorders including pheochromocytoma
        5) LV outflow obstruction
        6) Chronic high output states such as arteriovenous fistulas

3. Restrictive cardiomyopathy
    a. Endocardial fibroelastosis
    b. Amyloidosis
    c. Sarcoidosis
    d. Hemochromatosis
    e. Glycogen storage disease
    f. Metastatic disease
    g. Radiation

4. Arrhythmogenic right ventricular dysplasia
    a. Cardiomyopathy usually limited to the RV.
    b. Most common cause of arrhythmias originating from the RV
    c. Characterized by infiltration of the RV wall by fat and fibrous tissue.

    d. Typically involves the RV free wall but can extend into the ventricular septum and occasionally the LV.

  5. Uhl's anomaly (parchment RV)

    a. Marked hypoplasia or aplasia of the RV wall

    b. The epicardium and endocardium of the RV wall lie adjacent to each other with no intervening myocardial muscle.

    c. The interventricular septum is normal.

E. Afterload stress

  1. Hypertension

  2. Aortic stenosis

  3. Coarctation of aorta

References and suggested additional reading

1. Best CH, Taylor NB. *The Physiological Basis of Medical Practice, 8th ed.* Baltimore: Williams and Wilkins, 1966.
2. Breen JF. Imaging of the pericardium. J Thorac Imaging 2001; 16:47-54.
3. Carsky EW, Mauceri RA, Azimi F. The epicardial fat pad sign: analysis of frontal and lateral chest radiographs in patients with pericardial effusion. Radiology 1980; 137:303-308.
4. Carsky EW, Azimi F, Mauceri R. Epicardial fat pad sign in the diagnosis of pericardial effusion. JAMA 1980; 244:2762-2764.
5. Chen JTT, Putman CE, Hedlund LW, Damash NS, Roberts L. Widening of the subcarinal angle by pericardial effusion. AJR 1982; 139:883-887.
6. Elliott LP. The plain film approach utilizing differential diagnostic vascular anatomical sets. In: Taveras JM, Ferrucci JT eds. *Radiology: Diagnosis-Imaging-Intervention*. Volume 2, Chapter 13. Hagerstown: Lippincott, 1991.
7. Green CE, Elliott LP. The chest film in aortic valvular regurgitation. In: Taveras JM, Ferrucci JT eds. *Radiology: Diagnosis-Imaging-Intervention*. Volume 2, Chapter 46. Hagerstown: Lippincott, 1991.
8. Lane EJ Jr, Carsky EW. Epicardial fat: lateral plain film analysis in normals and pericardial effusion. Radiology 1968; 91:1-5.
9. Meszaros WT. *Cardiac Roentgenology: Plain Films and Angiocardiographic Findings*. Springfield: Charles C Thomas, 1969.
10. Miller SW. *Cardiac Radiology: The Requisites*. St. Louis: Mosby, 1996.
11. Proto AV, Speckman JM. The left lateral radiograph of the chest: part two. Med Radiogr Photogr 1980; 56:38-64.
12. Spindola-Franco H, Fish BG. *Radiology of the Heart: Cardiac Imaging in Infants, Children, and Adults*. New York: Springer-Verlag, 1985.
13. Tehranzadeh J, Kelley MJ. The differential density sign of pericardial effusion. Radiology 1979; 133:23-30.
14. Woodring JH. The lateral chest radiograph in the detection of pericardial effusion: a reevaluation. J Ky Med Assoc 1998; 96:218-224.

John H. Woodring, M.D.

Chapter 4.

Pulmonary venous hypertension

I. **Causes of pulmonary venous hypertension (PVH)**
   A. **Left ventricular (LV) failure of any cause (most common)**
   B. **Obstruction of the mitral valve (common)**
   C. **Mechanical obstruction proximal to the mitral valve (rare)**
   D. **Constrictive pericarditis (rare)**

II. **Theory of flow redistribution (cephalization) in PVH**
   A. **Normal distribution of pulmonary blood flow**
   1. As previously discussed, in the upright position there is a hydrostatic pressure gradient in the pulmonary arterial and venous circulations with hydrostatic pressure being greatest in the lung base and progressively diminishing as one moves toward the lung apex.
   2. The hydrostatic pressure gradient in the pulmonary arteries in the upright position ranges from approximately 25 mm Hg at the lung base to 5 mm Hg at the lung apex with a mean value of about 15 mm Hg. The hydrostatic pressure gradient in the pulmonary veins in the upright position ranges from 15 mm Hg at the base to –5 mm Hg at the lung apex with a mean value of about 5 mm Hg.
   3. Since intraalveolar pressure can be assumed to be constant over the vertical height of the lung, pulmonary blood flow in the upright position is directly related to the hydrostatic pressure gradient and is therefore greatest in the lung base and progressively decreases as one moves from the lung base to the apex.
   4. The pulmonary vasculature is easily distensible and there is a close relationship between vessel diameter and the amount of blood flowing through the vessels. Because of this, regional differences in hydrostatic pressure and pulmonary blood flow are reflected on the upright chest radiograph by regional differences in the size of the visible pulmonary arteries and veins, with the pulmonary vessels being larger in the lower lung zone than those in the upper lung zone. This results in a normal blood flow ratio (upper zone vessel size divided by lower zone vessel size) that is <1.
   B. **Physiologic effects of PVH**
   1. The pulmonary veins have no valves; therefore, pulmonary venous pressure directly reflects left atrial pressure (LAP). If the mitral valve is normal, LAP is equal to left ventricular end-diastolic pressure (LVEDP). Normally mean pulmonary venous pressure is 5 mm Hg.
   2. Fluid exchange across the pulmonary capillary wall obeys Starling's law and is related to differences in hydrostatic pressure and colloid-osmotic or

oncotic pressure between the plasma in the capillary and fluid in the surrounding lung interstitium. Fluid exchange across the capillary wall can be calculated by the following formula:

$$\text{Net fluid out} = K[(P_c - P_i) - (\pi_c - \pi_i)]$$

where K is a constant, P_c is the hydrostatic pressure in the capillary, P_i is the hydrostatic pressure of the free interstitial fluid, π_c is the oncotic pressure of plasma, and π_i is the oncotic pressure of the free interstitial fluid.

a. The force pushing fluid out of the capillary is the capillary hydrostatic pressure (P_c) minus the hydrostatic pressure of the interstitial fluid (P_i).

b. The force pulling fluid into the capillary is the osmotic pressure of the proteins in the plasma, the colloid-osmotic or oncotic pressure (π_c), minus the oncotic pressure of the interstitial fluid (π_i).

c. At the arterial end of the pulmonary capillary the mean hydrostatic pressure is approximately 15 mm Hg. Interstitial pressure is generally negative and is similar to intrapleural pressure which is about –3 mm Hg. This negative interstitial pressure helps keep the pulmonary microcirculation open by opposing the natural tendency of the small vessels to collapse. Therefore, at the arterial end of the pulmonary capillary there is a net hydrostatic pressure gradient of about 18 mm Hg driving fluid out of the capillary.

d. At the arterial end of the capillary plasma oncotic pressure is approximately 25 mm Hg and interstitial oncotic pressure is about 15 mm Hg. Therefore, at the arterial end of the pulmonary capillary there is a net oncotic pressure of 10 mm Hg pulling fluid back into the capillary. This force is 8 mm Hg less than the net hydrostatic force driving fluid out of the arterial end of the capillary; therefore, at the arterial end of the capillary there is a driving force of about 8 mm Hg that serves to filter fluid out of the capillary into the lung interstitium.

e. As blood flows through the capillary and fluid is lost into the surrounding interstitium, plasma oncotic pressure within the capillary progressively rises.

f. At the venous end of the capillary hydrostatic pressure is 5 mm Hg and interstitial pressure remains about –3 mm Hg resulting in a net hydrostatic pressure of only 8 mm Hg forcing fluid out of the capillary.

g. At the venous end of the capillary plasma oncotic pressure may be 30 mm Hg or higher. Interstitial oncotic pressure, however, remains about 15 mm Hg. Therefore, the net oncotic force pulling fluid back into the venous end of the capillary is about 15 mm Hg. This force, which is approximately 7 mm Hg greater than the hydrostatic force driving fluid out of the venous end of the capillary, results in absorption of fluid back into the capillary at the venous end.

h. Still, the net force pulling fluid back into the venous end of the capillary is slightly less than the net force driving fluid out of the arterial end of the capillary. Because of this difference in forces between the arterial and venous ends of the capillary there is a net loss of intravascular fluid into the lung interstitium as blood flows through the capillary. This excess interstitial fluid is normally carried away by the lymphatics.

3. As can be seen from the above example there is normally a delicate balance between hydrostatic pressure, oncotic pressure, and lymphatic flow. Increasing mean pulmonary venous pressure by only 7 mm Hg, to a mean pressure of 12 mm Hg, will eliminate the absorptive force which normally pulls interstitial fluid back into the venous end of the capillary. Eliminating the absorption of fluid into the venous end of the capillary increases the amount of fluid in the lung interstitium and raises interstitial hydrostatic pressure. When negative interstitial pressure is lost, the natural tendency of the pulmonary microcirculation to collapse is no longer opposed, and the pulmonary microcirculation constricts resulting in decreased blood flow. This sequence of events initiates the redistribution of pulmonary blood flow seen in PVH.

4. In the upright position, the constriction of the pulmonary microcirculation and slowing of blood flow from PVH typically occurs first in the lung bases because hydrostatic pressure in the pulmonary veins is already near the physiologic upper limits of normal there. This decrease in flow through the basilar microcirculation results in decreased flow through the larger pulmonary arteries and veins in the lung bases and is reflected on the upright PA chest radiograph by a decrease in size of the visible pulmonary arteries and veins in the lower lung zones. However, right ventricular output (5 L/min) is unchanged initially. Since blood flow in the bases is diminished, blood flow is diverted to the upper lung zones, which initially remain within physiologic normal limits. This is reflected by an increase in size of the visible arteries and veins in the upper lung zones on the upright PA chest radiograph. As a result of these changes pulmonary blood flow is inverted, or upside down, compared to normal.

5. The inversion of pulmonary blood flow in the lungs results in a blood flow ratio that is >1.
   a. This inversion of pulmonary blood flow secondary to PVH is called redistribution or cephalization of pulmonary blood flow.
   b. While many authors have focused only upon distention of the upper lung zone pulmonary veins in PVH, and while it is true that the upper lobe pulmonary veins are distended in PVH — occasionally to the point that the right hilar angle becomes flattened — the entire pulmonary circulation is inverted and the pulmonary arteries in the upper lung zones are enlarged as well.
   c. The enlarged upper lung zone vessels secondary to PVH have also been referred to as the "sergeant major's moustache" sign and the "touchdown" sign

III. Radiographic determination of presence of PVH in adults

A. Evaluation of upright PA radiographs

1. As in the determination of normal vascularity, in PVH one must compare vessels in the upper lung zone and lower lung zone that are of a similar order of branching from the central pulmonary artery.

2. The upper zone vessels used in this comparison are the segmental arteries and veins in the right upper lobe immediately above the right upper lobe bronchus. These vessels are encompassed by a small circle about 4-5 cm in diameter drawn on the frontal chest radiograph bounded inferiorly by the right upper lobe bronchus and medially by the superior vena cava.

3. The lower zone vessels used in this comparison are the segmental arteries and veins in the right lower lobe immediately below the descending branch of the right pulmonary artery. These vessels are encompassed by a small circle about 4-5 cm in diameter drawn on the frontal chest radiograph in the right cardiophrenic angle bounded inferiorly by the apex of the right hemidiaphragm and medially by the right atrium.

4. Compare the vascularity in the right and left lungs to see if the vascular changes are symmetrical.

5. If the upper zone vessels are visibly larger than the lower zone vessels on an upright PA chest radiograph, pulmonary blood flow is inverted and the patient has PVH.

6. Distention of the right superior pulmonary vein may cause the right hilar angle to become flattened.

7. The pulmonary artery-bronchus ratio (ABR) may also be helpful in detecting PVH.
   a. In PVH, the end-on upper lung zone pulmonary arteries are larger than their accompanying end-on bronchi (ABR >1).
   b. In the lower lung zone the end-on pulmonary arteries are smaller than their accompanying end-on bronchi (ABR <1).

B. The lateral chest radiograph

1. In PVH the pulmonary vessels in the lower lobes posterior to the heart are decreased in size and are noticeably smaller than the enlarged pulmonary vessels in the upper lobes above the hila on the lateral view.

2. This inversion of pulmonary flow on the lateral view has been called the "wineglass" sign — with the enlarged upper lung zone vessels representing the glass itself and the decreased lower lung zone vessels representing the stem of the glass.

C. Supine AP chest radiographs

1. In general, the discussion of the radiographic signs of PVH is limited to chest radiographs obtained with the patient in the upright position.

2. PVH is difficult to diagnose on supine AP portable chest radiographs because the gravitational hydrostatic pressure gradient in the lungs is eliminated in the supine position. Because of this, cephalization of pulmonary blood flow may not occur.

3. Still, many patients who are imaged in the supine position spend a good portion of time in an erect or semi-erect position. Because of this an inverted blood flow pattern may be seen in up to 73% of portable chest radiographs obtained in patients with PVH. In the remaining cases there is diffuse engorgement of the upper and lower lung zone vessels, with an ABR >1 in both the upper and lower lung zones, that mimics overcirculation of the pulmonary vascularity.

D. CT of the chest
1. Because CT is obtained in the supine position, cephalization of pulmonary blood flow is frequently absent. In fact, the gravitational gradient in the supine position is from back to front, rather than base to apex.
2. Cardiomegaly and chamber enlargement or hypertrophy are easily identified by CT, as are signs of right heart failure, which will be discussed later.
3. The ABR is very helpful in detecting PVH on CT scans of the chest. In PVH the pulmonary arteries become much larger than their accompanying bronchi with an ABR >1 in both the upper and lower lung zones. CT, however, cannot distinguish between PVH and overcirculation of the pulmonary vascularity because the ABR is >1 in both the upper and lower lung zones in the supine position in both conditions.

IV. Clinical measurement of pulmonary venous pressure
   A. Pulmonary capillary wedge pressure (PCWP)
   1. Is the clinical measure of pulmonary venous pressure
   2. Normal PCWP ranges from 2-12 mm Hg with a mean value of about 5 mm Hg.
   3. PCWP is measured in the supine position.
      a. In the supine position there is no gravitational hydrostatic gradient.
      b. Therefore, measured PCWP is normally uniform throughout both lungs.

V. Physiologic and radiographic stages of PVH
   A. Stage I — cephalization of pulmonary blood flow alone
   1. As discussed above, when mean pulmonary venous pressure reaches 12 mm Hg the absorptive force at the venous end of the capillary, which normally pulls interstitial fluid back into the capillary, is eliminated in the lung bases. This, in turn, increases the amount of fluid in the lung interstitium, eliminates the negative interstitial pressure that normally opposes the tendency of the pulmonary microcirculation to collapse, and results in constriction of the basilar pulmonary microcirculation with decreased blood flow through the lung bases. This sequence of events initiates cephalization of pulmonary blood flow.
   2. At the earliest stage, there is a slight reduction in size of the lower lung zone vessels and a slight increase in size of the upper lung zone vessels.

    a. This results in equal blood flow in the upper and lower lung zones and is referred to as "equalization" of pulmonary blood flow.

    b. Equalization of pulmonary blood flow is difficult to detect radiographically unless there are baseline normal radiographs for comparison, and this is frequently not the case.

3. By the time mean pulmonary venous pressure reaches 15 mm Hg cephalization of pulmonary blood flow is usually well-established and noticeable on upright PA chest radiographs. It is at this point that there is general agreement that the patient has PVH.

4. As mean pulmonary venous pressure approaches 20 mm Hg, distention of the right superior pulmonary vein may cause the right hilar angle to become flattened.

5. Cephalization of pulmonary blood flow takes time to develop and implies chronic PVH.

    a. Cephalization is common in chronic LV failure and mitral stenosis.

    b. Cephalization may be absent in acute LV failure.

B. Stage II — cephalization plus interstitial edema

1. Generally occurs with a mean pulmonary venous pressure from 20-25 mm Hg.

2. Once mean pulmonary venous pressure reaches 20 mm Hg, so much fluid is leaving the vascular space that the interstitium becomes visibly thickened radiographically.

    a. Radiographic signs

        1) Septal lines (Kerley A lines, Kerley B lines)

           a) Represent thickened interlobular septa

           b) Are seen in the periphery of the lung and extend to the pleural surface

           c) "A" lines occur at the lung apex, they are several centimeters in length, they are perpendicular to the pleural surface, and point downward toward the hila because of the arcuate shape of the lung apex.

           d) "B" lines occur at the lung base. They are about 1 cm in length, they are perpendicular to the pleural surface, and are parallel to the floor because of the vertical orientation of the interface between the chest wall and lung at the lung base.

        2) Reticulation

        3) Bronchial wall thickening

        4) Pulmonary vascular indistinctness or "haze"

        5) Fissural thickening

           a) The fissures, composed of two apposed layers of visceral pleura, are anatomically continuous with the interlobular septa.

           b) Thickening of the fissures is often an early sign of interstitial edema.

           c) Can be distinguished from intrafissural pleural effusion if the lateral and posterior costophrenic sulci remain sharp indicating that there is no pleural effusion present

     b. Although interstitial pulmonary edema is present at the microscopic level from the earliest stages of PVH, it is not diagnosed clinically and radiographically until one or more of the above signs appear on the chest radiograph.

C. **Stage III — cephalization plus interstitial and alveolar edema**
1. Occurs with a mean pulmonary venous pressure above 25 mm Hg
2. Interstitial edema spills out of the interstitium into the alveoli causing alveolar edema.
3. Alveolar pulmonary edema secondary to PVH is usually infrahilar and basilar in distribution and extends outward toward the lateral costophrenic sulci.
4. Cephalization and interstitial edema are still present but may be obscured by the alveolar edema.

D. **Stage IV — pulmonary hemosiderosis and ossification**
1. Occurs in long-standing severe PVH
2. Usually seen only in mitral stenosis but rarely may be seen in some cases of left atrial myxoma

E. **Variations in the correlation between signs of PVH and mean pulmonary venous pressure**
1. If plasma oncotic pressure is decreased (hypoalbuminemia, hypoproteinemia) the radiographic signs of PVH will occur at lower levels of mean pulmonary venous pressure.
2. If PVH is long-standing, pulmonary lymphatic flow will increase significantly and many of the signs of PVH, particularly those of interstitial and alveolar edema, will not occur until the mean pulmonary venous pressure is significantly higher than usual.

VI. **Radiographic determination of presence of PVH in infants**
A. **Cephalization of blood flow is usually absent.**
1. Because of the small size of the thorax in infants, and the fact that infants live in a supine position and are imaged in the supine position, a gravitational hydrostatic pressure gradient does not develop in infants' lungs. As a result, cephalization of pulmonary blood flow usually does not occur.
2. Cephalization may be seen in some older infants and children after they are able to sit or stand in an erect position.

B. **PVH is usually diagnosed based on the presence of signs of interstitial or alveolar pulmonary edema.**
1. Reticulation is one of the better radiographic signs of interstitial edema in infants and is, therefore, one of the better radiographic signs of PVH.
2. Other signs of interstitial edema including Kerley A and B lines, vascular indistinctness, and fissural thickening can also occur.
3. If PVH is severe there may be bilateral perihilar alveolar edema.

VII. Pleural effusion in PVH
A. Production of pleural effusion
1. Increased pressure at the venous end of the pulmonary capillaries causes the development of interstitial edema which tracks out along the interlobular septa to the visceral pleural surface of the lung and leaks directly into the pleural space.
2. The superimposition of right heart failure raises systemic venous pressure and impairs parietal pleural lymphatic drainage, which may also contribute to the accumulation of pleural effusion.
3. Pleural effusion can develop during any stage of PVH, since interstitial edema is present from the earliest stages of PVH.

B. Radiographic distribution of pleural effusion
1. Bilateral pleural effusion (73%)
   a. Right = left (30%)
   b. Right > left (21%)
   c. Left > right (22%)
2. Unilateral pleural effusion (28%)
   a. Right only (15%)
   b. Left only (13%)

VIII. Physiologic and radiographic spectrum of LV failure
A. Acute (first episode) LV failure
1. Since cephalization of pulmonary blood flow takes time to develop and implies chronic PVH, noticeable cephalization of pulmonary blood flow may be absent during the first episode of LV failure. Rather, many cases show only equalization of pulmonary blood flow.
2. The heart size is often normal but may be enlarged.
3. Systemic venous pressure and blood volume are normal; therefore the vascular pedicle is normal.
4. There are frequently signs of interstitial pulmonary edema.
5. Alveolar edema, when present, is infrahilar and basilar in distribution.
6. Pleural effusion may be present but is usually of small size.

B. Cardiogenic shock
1. Usually due to acute massive myocardial infarction (MI)
2. There is acute, severe elevation of pulmonary venous pressure
3. Since the patient often goes rapidly from a normal physiologic state prior to the MI to severe PVH immediately after the MI, there may not be time for cephalization of pulmonary blood flow to develop. As in acute (first episode) LV failure, many cases show only equalization of pulmonary blood flow.
4. The heart size may be normal or enlarged.
5. Systemic venous pressure and blood volume are initially normal; therefore the vascular pedicle is also initially normal.
6. Interstitial pulmonary edema is almost always present; however, radiographic signs of interstitial edema may be obscured by alveolar edema.

7. Central perihilar pulmonary alveolar edema with a "butterfly" or "bat's-wing" distribution is typical.
8. Pleural effusion may be present but is usually of small size initially.

C. Chronic LV failure
1. Cephalization of pulmonary blood flow is almost always present.
2. LV enlargement is common — may have secondary left atrial enlargement from dilatation of the mitral annulus
3. Signs of interstitial pulmonary edema are common.
4. Alveolar edema, when present, is infrahilar and basilar in distribution.
5. Pleural effusion is common.
6. The vascular pedicle is normal at first; however, chronic LV failure often causes decreased renal perfusion with associated plasma volume overload and widening of the vascular pedicle to the right.

D. Biventricular failure
1. Long-standing left heart failure can eventually cause the right heart to fail as well. In fact, chronic left heart failure is the most common cause of right heart failure.
2. Often have enlargement of all four cardiac chambers
3. Cephalization of pulmonary blood flow is almost always present; however, severe right heart failure can diminish, and occasionally eliminate, the manifestations of left heart failure.
4. Signs of interstitial pulmonary edema are common.
5. Alveolar edema, when present, is infrahilar and basilar in distribution.
6. Pleural effusion is common.
7. Right heart failure leads to systemic venous hypertension.
   a. Radiographic manifestations of systemic venous hypertension
      1) Widened vascular pedicle with enlargement of the superior vena cava and azygos vein
         a) Can also be seen in chronic LV failure with decreased renal perfusion and associated plasma volume overload
         b) The other radiographic signs of right heart failure listed below, however, usually do not occur from simple plasma volume overload alone and favor a diagnosis of biventricular failure.
      2) Posterior bulge of the suprahepatic portion of the inferior vena cava (IVC) on lateral chest radiograph
         a) The posterior border of the IVC is normally concave.
         b) In right heart failure the posterior border of the IVC may become convex.
      3) On CT scans of the abdomen there will be passive congestion of the liver (mottled hepatic parenchymal enhancement and reflux of contrast-enhanced blood from the right atrium into the IVC and hepatic veins), enlargement of the IVC and hepatic veins, and often some degree of ascites.
      4) Thickened chest wall — is a good sign of right heart failure but is difficult to diagnose unless there are baseline films for comparison demonstrating the normal thickness of the patient's chest wall.

E. Acute LV decompensation superimposed on chronic LV failure

1. The superimposition of acute LV decompensation on chronic LV failure, such as occurs when a patient with chronic LV failure suffers an acute MI, can produce a somewhat unusual radiographic picture.
2. Cephalization of pulmonary blood flow and other signs of chronic LV failure are present.
3. Because of the chronic LV failure and resultant decrease in pulmonary blood flow at the lung bases, the lung bases are relatively protected from developing acute alveolar edema from the superimposed acute and often severe elevation of pulmonary venous pressure. Instead of developing basilar or perihilar alveolar pulmonary edema, these patients may develop bilateral upper lobe alveolar pulmonary edema.

F. LV failure superimposed on pulmonary emphysema

1. The presence of emphysema greatly alters the cephalization of pulmonary blood flow secondary to LV failure due to destruction of the pulmonary vascular tree in the emphysematous portions of the lungs.
2. In basilar predominant emphysema blood flow is typically shunted to the more normal upper lung zones, a situation referred to as "organic redistribution" to indicate that the increased pulmonary blood flow in the upper lung zones is due to the primary pulmonary disease rather than PVH. When LV failure occurs in these individuals there is a further decrease in blood flow to the oligemic lower lung zones and a further increase in blood flow to the upper lung zones that can be mistaken for worsening of the basilar emphysema.
3. In mild apical predominant emphysema, cephalization of pulmonary blood flow from superimposed LV failure may cause a slight increase in size of the upper lobe vessels and return the chest radiograph to an almost normal appearance.
4. In severe apical predominant emphysema, there may be so much destruction of the upper lobe vasculature that no significant cephalization of pulmonary blood flow is possible. In this circumstance PVH may be evidenced as further prominence of the basilar vessels.
5. In patients with patchy, random areas of emphysema, cephalization of blood flow secondary to LV failure may fill in the oligemic areas making the chest radiograph appear almost normal, or may result in marked prominence of the pulmonary vascularity in the more normal areas of lung between the emphysematous areas.
6. Therefore, although cephalization of pulmonary blood flow will occur in most cases, the radiographic picture is often atypical, consisting of markedly prominent upper lung zone vessels, markedly prominent lower lung zone vessels, randomly prominent vessels between oligemic areas, or a near normal vascular appearance.
7. Clues to the diagnosis of LV failure superimposed on emphysema
   a. Indistinctness of the pulmonary vessels secondary to perivascular interstitial edema and Kerley B lines are cardinal features of LV failure

and help in the assessment of atypical cephalization patterns in patients with emphysema.

b. Interstitial edema reduces lung compliance and returns lung volume toward normal. Previous radiographs may show flattened, low-positioned hemidiaphragms with rounded costophrenic angles; however, when LV failure is superimposed, the hemidiaphragms often assume a normal or high position with sharp costophrenic angles.

c. In severe emphysema the cardiothoracic ratio (CTR) is often reduced and may normally be in the range of 25-30%. Since the patient's heart can enlarge considerably, and the CTR still be within normal limits for the population as a whole, any interval increase in size of the cardiac silhouette from baseline radiographs should be viewed as supportive evidence of possible LV failure.

d. It is also helpful to remember that pleural effusion is not a sign of emphysema, and when pleural effusion is encountered in a patient with known emphysema, superimposed LV failure should be considered as a possibility.

IX. Comparison of clinical and radiographic diagnosis of heart failure
A. Clinical diagnosis of heart failure
1. The clinical diagnosis of heart failure is usually made on the basis of an appropriate clinical history (breathlessness, fatigue, and exercise intolerance) coupled with physical findings of circulatory congestion including tachypnea, jugular venous distention, gallop rhythm, rales, peripheral edema, and ascites.
2. In early LV failure physical findings are initially limited to tachypnea and gallop rhythm. Rales may be heard if macroscopic pulmonary edema develops. As LV failure becomes chronic, signs of fluid retention (pedal edema) and superimposed right heart failure (jugular venous distention, hepatomegaly, hepatic tenderness, elevated liver function tests, and ascites) predominate.

B. Radiographic diagnosis of heart failure
1. Since cephalization of pulmonary blood flow from PVH develops earlier than many of the clinical signs of LV failure, the chest radiograph is a better detector of early LV failure than the clinical examination.
2. The radiographic signs of right heart failure consist of widening of the vascular pedicle with enlargement of the superior vena cava and azygos vein, a posterior bulge of the suprahepatic portion of the IVC on the lateral chest radiograph, thickening of the chest wall, and CT evidence of passive congestion of the liver, enlargement of the abdominal portion of the IVC and hepatic veins, and ascites.
3. Since widening of the vascular pedicle can occur from chronic LV failure with plasma volume overload from decreased renal perfusion without superimposed right heart failure, and the other radiographic signs of right heart failure may either be absent occasionally or require CT of the

41

abdomen to identify, the clinical examination is a better detector of right heart failure than the chest radiograph.

X. Differential diagnosis of LV failure — mnemonic = "PICA"
   A. Preload stress
       1. Conditions that produce preload stress on the LV
          a. Aortic valvular regurgitation
             1) Cephalization of pulmonary blood flow and other signs of PVH may be present.
             2) Left ventricular enlargement (LVE) characteristic
             3) Left atrial enlargement (LAE) may develop late from dilatation of mitral annulus.
             4) Enlarged aorta, particularly the aortic arch (aortic knob) and descending aorta
          b. Mitral valvular regurgitation
             1) Cephalization of pulmonary blood flow and other signs of PVH may be present.
             2) Usually have both LAE and LVE, may also have enlarged left atrial appendage (LAA)
             3) LVE absent in pure mitral stenosis
   B. Ischemia
       1. Coronary artery disease
          a. Cephalization of pulmonary blood flow and other signs of PVH may be present.
          b. May have LVE and secondary LAE
          c. Look specifically for coronary artery calcification, calcified MIs, and calcified LV aneurysms.
       2. Complications of MI
          a. LV aneurysm
             1) True aneurysm
                a) Contains thinned, dysfunctional LV wall
                b) Usually involves LV apex and is anteriorly located on lateral view
                c) May have thin peripheral calcification
                d) Associated with arrhythmias
                e) Low incidence of rupture
             2) False aneurysm
                a) Represents post-MI rupture of LV contained only by pericardium
                b) Posteriorly located on lateral view
                c) High incidence of rupture and sudden death
          b. Papillary muscle dysfunction with mitral regurgitation
             1) May be due to ischemia of the papillary muscles or adjacent LV free wall

       2) May develop pulmonary edema predominantly in the right upper lobe from a jet of regurgitant flow into the right superior pulmonary vein which is directly opposite the mitral valve

  c. Rupture of the papillary muscles

       1) Complete rupture of the papillary muscles is rare and often results in death.

       2) Presents as sudden onset of severe alveolar pulmonary edema 2-7 days after acute MI

       3) Rupture of the posteromedial papillary muscle is more common than rupture of the anterolateral papillary muscle.

  d. Ruptured interventricular septum with acquired ventricular septal defect

       1) May develop acute pattern of overcirculation associated with sudden onset of alveolar pulmonary edema.

       2) Pulmonary edema is usually not as severe as that seen with rupture of the papillary muscles.

       3) Rare

  3. Anomalous left coronary artery

    a. The left coronary artery arises from the main pulmonary artery (MPA) rather than the aorta.

    b. Infants experience angina with crying and diaphoresis as pulmonary artery pressures fall after birth.

    c. The LV and left atrium (LA) are dilated resulting in a markedly enlarged cardiac silhouette usually associated with severe reticulation or alveolar pulmonary edema. Cephalization of pulmonary blood flow is usually not present.

    d. Treatment — surgical attachment of the anomalous left coronary artery to the aorta.

C. Cardiomyopathy

  1. Dilated cardiomyopathy

    a. Cephalization of pulmonary blood flow and other signs of PVH may be present.

    b. The cardiac silhouette is typically globally enlarged.

  2. Hypertrophic cardiomyopathy

    a. Cephalization of pulmonary blood flow and other signs of PVH may be present.

    b. The cardiac silhouette may be normal in size but there is usually some degree of cardiomegaly due to the ventricular hypertrophy.

  3. Restrictive cardiomyopathy

    a. Cephalization of pulmonary blood flow and other signs of PVH may be present.

    b. Initially the cardiac silhouette is normal in size but as the disease progresses cardiomegaly may develop from a combination of chamber enlargement and hypertrophy.

D. Afterload stress

  1. Hypertension

    a. Cephalization of pulmonary blood flow and other signs of PVH may be present.

    b. LVE common

    c. Enlarged aortic knob and descending aorta

2. Aortic stenosis (AS) in adults — including both acquired AS (rheumatic, degenerative) and congenital bicuspid AS

    a. Cephalization of pulmonary blood flow and other signs of PVH may be present.

    b. LVE common

    c. Enlarged ascending aorta

    d. Look for calcification in valve on lateral view

3. AS in infants — congenital AS

    a. Accounts for 5% of cases of congenital heart disease

    b. Classification

        1) Valvular — 75%

            a) Usually due to bicuspid aortic valve

            b) Can also be unicuspid (unicommissural) or tricuspid with severe dysplasia

            c) When severe, can be part of the hypoplastic left heart syndrome

                i) Often associated with small, hypoplastic LV

                ii) Presents with severe congestive heart failure immediately after birth

        2) Subvalvular — 20%

            a) Obstructive hypertrophic cardiomyopathy — formerly idiopathic hypertrophic subaortic stenosis (IHSS)

            b) Discrete membranous subaortic stenosis — due to a discrete fibrous membrane or web less than 1 mm thick

            c) Subaortic fibrous tunnel — diffuse fibrous or fibromuscular thickening of the left ventricular outflow tract resulting in a tunnel-like obstruction

            d) Shone syndrome — association of coarctation of the aorta, discrete membranous or fibromuscular subaortic stenosis, parachute deformity of the mitral valve, and supravalvular ring of the LA

                i) Parachute deformity of the mitral valve results when all of the chordae tendineae converge on a single papillary muscle — results in mitral stenosis (MS)

                ii) Supravalvular ring of the LA is a circumferential ridge of connective tissue protruding into the inlet of the mitral valve at the base of the LA — can also cause MS

        3) Supravalvular — 5%

            a) Focal or diffuse narrowing of the ascending aorta just above the coronary arteries

                i) Hourglass deformity of the ascending aorta — 65%

                ii) Simple fibrous diaphragm of the supravalvular aorta — 10%

        iii) Diffuse tubular narrowing (hypoplasia) of the ascending aorta beginning just above the coronary arteries — 25%

    b) Usually part of William's syndrome

        i) Mental retardation

        ii) Hypercalcemia

        iii) Elfin facies

        iv) Supravalvular AS

        v) Supravalvular/peripheral pulmonic stenosis

  c. Severe AS in infants usually presents with cardiomegaly and signs of PVH, particularly severe reticulation from interstitial edema.

4. Coarctation of the aorta

  a. Cephalization of pulmonary blood flow and other signs of PVH may be present.

  b. LVE common

  c. Look for primary and collateral signs of coarctation

5. Hypoplastic left heart syndrome

  a. Accounts for 2% of cases of congenital heart disease

  b. Consists of a group of lesions characterized by one or more of the following:

    1) Aortic atresia or severe AS — 95%

    2) Hypoplastic ascending aorta and aortic arch including interruption of the aortic arch

    3) Small hypoplastic LV

    4) Small hypoplastic LA

    5) Mitral atresia or severe MS — 25%

  c. Usually has an atrial septal defect (ASD) and a patent ductus arteriosus (PDA) that allow blood to bypass the hypoplastic left heart.

    1) A small patent foramen of ovale or other type of ASD allows blood to flow from the LA to the right atrium.

    2) A large PDA allows for blood to flow from the pulmonary artery to the systemic circulation.

  d. Clinical manifestations consist of congestive heart failure in the first week of life occasionally accompanied by mild cyanosis.

    1) Hypoplastic left heart syndrome is the most common cause of congestive heart failure in the first week of life.

    2) It is the most common cause of neonatal cardiac death, accounting for 25% of all cardiac deaths in the first week of life.

  e. Radiographic manifestations usually consist of cardiomegaly, enlargement of the MPA, and signs of PVH including interstitial or alveolar pulmonary edema, although overcirculation of the pulmonary vascularity may occasionally be present.

  f. Treatment

    1) Palliative treatment — Norwood procedure

      a) Has two variations

      b) Systemic blood flow is established from the right ventricle (RV) by placement of a conduit from the RV free wall or proximal

MPA to the aorta. Pulmonary blood flow is limited either by pulmonary artery banding and ligation of the PDA or performance of a Blalock-Taussig procedure (anastomosis of the subclavian artery on the side opposite the aortic arch to the pulmonary artery) to supply blood to the pulmonary arteries.

c) Systemic blood flow is established by direct anastomosis of the proximal MPA and ascending aorta. Pulmonary vasculature is protected by a Blalock-Taussig shunt.

2) Definitive treatment — heart transplant

XI. PVH secondary to obstruction of the mitral valve
A. Rheumatic mitral stenosis (MS)

1. Clinical and radiographic manifestations of MS usually occur 10-20 years after an episode of acute rheumatic carditis but can develop earlier.
2. Almost always secondary to rheumatic carditis; however, about 60% of patients do not give a history of prior rheumatic fever.
3. More common in females than males. M:F = 1:8
4. Cephalization of pulmonary blood flow is a cardinal feature of MS.
5. The heart typically has a characteristic configuration which consists of the following:
   a. LAE with a "double-density" behind the right side of the heart with an oblique LA dimension >7 cm.
      1) It should be noted that a normal LA can produce a "double-density" behind the right side of the heart.
      2) The oblique LA dimension is extremely helpful in deciding whether this "double-density" represents a normal or enlarged LA.
      3) The double-density of an enlarged LA behind the right heart has also been called the "baseball" sign.
      4) LAE may cause widening of the carinal angle on the PA chest radiograph.
      5) On the lateral chest radiograph the upper portion of the posterior heart border below the carina may bulge posteriorly and also displace the left mainstem bronchus posteriorly.
      6) On barium swallow an enlarged LA displaces the subcarinal portion of the esophagus posteriorly.
   b. Enlarged LAA
      1) To those familiar with downhill skiing, the left heart border usually has 3 moguls or bumps: the aortic arch, the MPA, and the LV apex.
      2) Enlargement of the LAA produces a fourth mogul below the MPA segment at the base of the heart.
      3) The enlarged LAA produces the classic "mitral configuration" of the heart.
      4) Open mitral commissurotomy may be done as an early surgical treatment for MS. When this is done the LA is entered through the LAA and at the end of the procedure the LAA is resected to reduce

the risk of post-operative LA thrombosis. These patients will no longer have an enlarged LAA on the chest radiograph — look closely for signs of prior left thoracotomy as a clue.

   c. Normal-sized LV
      1) In pure MS the LV is of normal size.
      2) LVE indicates associated mitral regurgitation.

6. Signs of interstitial edema are common, and alveolar edema may also occur. Because of the chronicity of PVH in MS, pulmonary lymphatic flow increases considerably and radiographic signs of interstitial and alveolar edema usually do not occur until pulmonary venous pressure is considerably higher than the usual values expected to cause interstitial and alveolar edema.

7. When MS is long-standing and severe there is often pulmonary arterial hypertension (PAH) with enlargement of the central pulmonary arteries, MPA, and evidence of right heart failure.

8. Associated findings in rheumatic MS
   a. Aortic and tricuspid valve disease may also be present
   b. LA calcification
      1) Is rare
      2) Occurs in LA endocardium and subendocardium
      3) Is presumably caused by rheumatic endocarditis
      4) 87% have associated LA thrombosis; however, most cases of LA thrombosis do not have LA wall calcification
   c. LA thrombus
      1) Produces a filling defect in the LA.
         a) Can be mistaken for LA myxoma
         b) LA myxoma is usually attached to the atrial septum.
         c) LA thrombus is usually attached to the posterior wall of the left atrium.
      2) Can cause systemic arterial embolization
   d. Mitral valve calcification
      1) Occurs in 10% of cases
      2) On the PA chest radiograph projects to the left of the spine
      3) On the lateral view usually seen in the posterior third of the cardiac silhouette below a line drawn between the end-on left mainstem bronchus and the anterior cardiophrenic angle
      4) May not be visible on plain films — non-contrast enhanced CT is superior to plain films in detecting mitral valve calcification
   e. Pulmonary hemosiderosis
      1) Develops from chronic PVH and repeated areas of small intra-alveolar pulmonary hemorrhage
      2) Typically produces a micronodular pattern in the lungs with small nodules ranging from 1-5 mm in diameter.
      3) Is seen radiographically in 5% of cases of MS but the pathologic incidence is much higher
   f. Pulmonary ossification

       1) Also develops from chronic PVH and chronic pulmonary edema

       2) Is less common than pulmonary hemosiderosis and is unrelated to pulmonary hemosiderosis

       3) Most commonly seen in MS but can also occur from LA myxoma with obstruction of the mitral valve

       4) Multiple areas of actual ossification occur in the lung producing ossified pulmonary nodules ranging in size from 2-10 mm in diameter.

       5) Mid and lower lung zone predominant

       6) Resembles multiple calcifications which may occasionally be seen in histoplasmosis

    g. Pulmonary venous varix may be present if there is associated mitral regurgitation.

B. Lutembacher's syndrome

    1. Coincidental association of congenital ASD and rheumatic MS (extremely rare)

    2. The radiographic manifestations are variable and depend upon the size of the ASD

      a. The right side of the heart is almost always enlarged.

      b. If the ASD is small, the LA will be enlarged, there may be cephalization of pulmonary blood flow and other signs of PVH including vascular indistinctness and basilar predominant septal edema.

      c. If the ASD is large, the LA is decompressed by the ASD and may be of normal size. In these cases the chest radiograph usually shows marked overcirculation of the pulmonary vascularity, as one would expect in a large ASD. Signs of interstitial edema from PVH including vascular indistinctness and basilar predominant septal edema may also occur.

      d. The MPA is usually markedly enlarged.

      e. PAH and right heart failure are common.

      f. May have calcification of the mitral valve

    3. The uncommon association of a "mitral configuration" and overcirculation of the pulmonary vascularity is suggestive of the diagnosis.

C. Left atrial myxoma

    1. Usually arises from the atrial septum

    2. May obstruct the mitral valve and cause typical clinical and radiographic signs of MS

    3. The first heart sound may be prominent and a diastolic "plop" or "thump" may be heard on auscultation.

D. Hypoplastic left heart syndrome with mitral atresia or stenosis

E. Other causes of MS

    1. Active infective endocarditis

    2. Massive annular calcification

    3. Systemic lupus erythematosus

    4. Rheumatoid arthritis

5. Carcinoid tumor
6. Tumor invasion of the pulmonary veins with extension into the LA
7. Methysergide therapy
8. Hunter-Hurler syndrome
9. Fabry's disease
10. Whipple's disease

XII. PVH secondary to obstruction proximal to the mitral valve
   A. Stenosis or atresia of the pulmonary veins
   1. May be congenital or acquired
   2. Acquired causes include fibrosing mediastinitis, tumor invasion, and constrictive pericarditis.
   3. May involve all or only one pulmonary vein
   4. Radiographic findings:
      a. Redistribution of pulmonary blood flow is usually absent.
      b. Typically characterized by severe reticulation
      c. The MPA segment may be prominent.
      d. There may be right ventricular enlargement (RVE).
      e. Reticulation may be unilateral if the stenosis is unilateral.
   B. Cor triatriatum
   1. Results from failure of incorporation of the common pulmonary vein into the LA
   2. The common pulmonary vein persists as a separate chamber that communicates with the LA through a small opening which is usually stenotic.
   3. Radiographic findings:
      a. Redistribution of pulmonary blood flow is absent.
      b. Characterized by severe reticulation
      c. The MPA is prominent.
      d. There is RVE.
   C. Totally anomalous pulmonary venous return (obstructed)
   1. Occurs in total infracardiac connection of the pulmonary veins to the portal vein or ductus venosus
   2. The anomalous right and left pulmonary veins converge into a single venous trunk that follows the esophagus through the esophageal hiatus in the diaphragm into the abdomen.
   3. The anomalous venous trunk is obstructed as it courses through the esophageal hiatus.
   4. The infants are cyanotic, and the cyanosis typically worsens during feeding because of additional compression of the venous trunk by the esophagus as it courses through the esophageal hiatus.
   5. Radiographic findings:
      a. Characterized by severe reticulation and Kerley B line formation
      b. Redistribution of pulmonary blood flow is absent.
      c. Cardiac size is usually normal.

D. Pulmonary veno-occlusive disease (PVOD)

1. Typically seen in children and adolescents but can occur in adults
2. Many cases may be due to post-viral obliteration of the pulmonary veins. Ingested substances have also been implicated including bleomycin, mitomycin, herbal bush teas, and oral contraceptives. PVOD may also occur after bone marrow transplantation and radiation therapy.
3. Characterized by obliteration of the lumen of small pulmonary veins and venules by intimal fibrous tissue. The larger pulmonary veins are spared.
4. Radiographic findings:
   a. Signs of PAH are present
      1) Enlarged central pulmonary arteries with pruning of the peripheral pulmonary arteries
      2) Enlarged MPA
      3) RVE
   b. There is no cephalization of pulmonary blood flow because the pulmonary veins are obliterated.
   c. There is no LAE or LVE
   d. Typically there is interstitial edema, manifested by reticulation and Kerley B line formation, and there may be alveolar edema as well.
   e. The combination of PAH and pulmonary edema without LAE, LVE, and cephalization of pulmonary blood flow is suggestive of the diagnosis.

XIII. Constrictive pericarditis

A. Rare cause of PVH
B. May occur if there is significant compression of LV or obstruction of the pulmonary veins

References and suggested additional reading

1. Best CH, Taylor NB. *The Physiological Basis of Medical Practice, 8th ed.* Baltimore: Williams and Wilkins, 1966.
2. Dähnert W. *Radiology Review Manual 4th ed.* Philadelphia: Lippincott Williams & Wilkins, 2000.
3. Diem K, Lentner C eds. *Documenta Geigy Scientific Tables 7th ed.* Basle: Ciba-Geigy, 1970.
4. Ewald GA, McKenzie CR eds. *Manual of Medical Therapeutics 28th ed.* Boston: Little, Brown, 1995.
5. Fuster V, Alexander RW, O'Rourke RA, Roberts R, King SB III, Wellens HJJ. *Hurst's The Heart 10th ed.* New York: McGraw-Hill, 2001.
6. Grossman W, Baim DS. Diagnostic cardiac catheterization and angiography. In: Isselbacher KJ, Braunwald E, Wilson JD, Martin JB, Fauci AS, Kasper DL eds. *Harrison's Principles of Internal Medicine 13th ed.* New York: McGraw-Hill, 1994.
7. Keele CA, Neil E. *Samson Wright's Applied Physiology 12th ed.* London: Oxford University Press, 1971.

8. Koo BC, Woldenberg LS, Kim K-T. Pulmonary vein tumor thrombosis and left atrial extension in lung carcinoma. J Comput Tomogr 1984; 8:331-336.

9. Lang P, Norwood WI. Hemodynamic assessment after palliative surgery for hypoplastic left heart syndrome. Circulation 1983; 68:104-108.

10. Meszaros WT. *Cardiac Roentgenology: Plain Films and Angiocardiographic Findings*. Springfield: Charles C Thomas, 1969.

11. Miller SW. *Cardiac Radiology: The Requisites*. St. Louis: Mosby, 1996.

12. Milne ENC, Bass H. Roentgenologic and functional analysis of combined chronic obstructive pulmonary disease and congestive cardiac failure. Invest Radiol 1969; 4:129-147.

13. Milne ENC, Pistolesi M. *Reading the Chest Radiograph: a Physiologic Approach*. St. Louis: Mosby, 1993.

14. Milne ENC, Pistolesi M, Miniati M, Giutini C. The vascular pedicle of the heart and the vena azygos. Part 1: the normal subject. Radiology 1984; 152:1-8.

15. Müller NL, Fraser RS, Colman NC, Paré PD. *Radiologic Diagnosis of Diseases of the Chest*. Philadelphia: Saunders, 2001.

16. Pistolesi M, Milne ENC, Miniati M, Giutini C. The vascular pedicle of the heart and the vena azygos. Part 2: acquired heart disease. Radiology 1984; 152:9-17.

17. Simon M. The pulmonary vessels: their hemodynamic evaluation using routine radiographs. Radiol Clin North Am 1963; 1:363-376.

18. Simon M. Physiologic considerations in radiology of the pulmonary vasculature. In: Abrams HL ed. *Abrams Angiography Vascular and Interventional Radiology 3rd ed*. Boston: Little, Brown, 1983:783-802.

19. West JB. *Respiratory physiology—the Essentials*. Baltimore: Williams and Wilkins, 1974.

20. Woodring JH. Pulmonary artery-bronchus ratios in patients with normal lungs, pulmonary vascular plethora, and congestive heart failure. Radiology 1991; 179:115-122.

21. Woodring JH, West JW. CT of aortic and mitral valve calcification. J Ky Med Assoc 1989; 87:177-180.

John H. Woodring, M.D.

Chapter 5.

Overcirculation of the pulmonary vascularity

I. **Causes of overcirculation of the pulmonary vascularity**
   A. **Plasma volume overload**
   B. **High-flow states**
   C. **Acyanotic left-to-right (L-R) shunts**
   D. **Cyanotic admixture lesions**

II. **Physiologic considerations in overcirculation of the pulmonary vascularity (pulmonary vascular plethora)**
   A. **Increased pulmonary blood flow and pulmonary blood volume**
      1. Any increase in pulmonary blood flow is generally accompanied by a simultaneous decrease in transit time through the lungs and an increase in pulmonary blood volume. Because of this, increases in pulmonary blood flow and pulmonary blood volume are inseparable and the radiologic signs within the lungs for the two conditions are identical. Likewise, any increase in pulmonary blood volume is also usually accompanied by an increase in pulmonary blood flow.
      2. Because of the easy distensibility of the pulmonary circulation there is a close relationship between pulmonary blood flow and vessel diameter. This flow-caliber relationship can be described by the formula: flow (cc/min) = D^2 X 10, where D is the vessel diameter.
      3. Assuming that the mitral valve and left ventricle (LV) are normal, and there is no obstruction to flow through the left heart, the physiologic alterations that occur in the lungs secondary to increased pulmonary blood flow and pulmonary blood volume are not gravitationally dependent. Rather, increased pulmonary blood flow and pulmonary blood volume are manifested by a diffuse increase in the caliber of the pulmonary arteries and veins in the upper and lower lung zones of both lungs. This results in a "balanced" overcirculation pattern with a 1:1 blood flow ratio.
      4. Furthermore, hydrostatic pressures in the pulmonary circulation are usually not elevated sufficiently to cause interstitial or alveolar edema until the degree of pulmonary vascular plethora is severe.
   B. **Systemic blood volume**
      1. Increased pulmonary blood flow and blood volume may occur as part of an overall increase in circulating blood volume that is also associated with increased systemic blood volume.
         a. This situation typically occurs in iatrogenic fluid overload and renal failure.
         b. The vascular pedicle is widened to the right.
            1) The superior vena cava (SVC) and azygos vein are enlarged.

> 2) The right atrium (RA) is increased in size compared to baseline radiographs.

2. Conversely, increased pulmonary blood flow and blood volume can be associated with diminished systemic blood volume.
   a. This situation typically occurs with moderate-to-large acyanotic L-R intracardiac shunts.
   b. The vascular pedicle is small.
3. While the radiographic manifestations of pulmonary vascular plethora in the lungs are the same regardless of the cause, assessment of the vascular pedicle is very helpful in determining the etiology of the high-flow state.

III. Radiographic signs of overcirculation of the pulmonary vascularity
A. Overcirculation of the pulmonary vascularity in adults
1. Findings on upright posteroanterior (PA) chest radiographs
   a. Pulmonary blood flow and volume are increased resulting in an increase in the size of the pulmonary arteries and pulmonary veins in both the upper and lower lung zones resulting in a "balanced" overcirculation pattern with a 1:1 blood flow ratio.
   b. Blood flow should be symmetrical in both lungs.
   c. The end-on pulmonary arteries are larger than their accompanying end-on bronchi in both the upper and lower lung zones resulting in a pulmonary artery-bronchus ratio (ABR) >1 in both the upper and lower zones.
   d. Vessels seen through the liver are too big.
2. Findings on the lateral chest radiograph
   a. The pulmonary arteries and veins in the lower lung zones behind the heart are enlarged, as are the pulmonary arteries and veins above the hila.
   b. Again, a "balanced" blood flow pattern with a 1:1 blood flow ratio is encountered.
3. Findings on supine anteroposterior (AP) chest radiographs
   a. As previously stated, the determination of pulmonary vascular patterns is generally based on upright PA and lateral chest radiographs rather than on supine AP views that may be obtained with portable technique.
   b. In pulmonary vascular plethora supine AP chest radiographs show an increase in the size of the pulmonary arteries and pulmonary veins in both the upper and lower lung zones resulting in a "balanced" overcirculation pattern with a 1:1 blood flow ratio. The end-on pulmonary arteries are larger than their accompanying end-on bronchi in both the upper and lower lung zones resulting in an ABR >1 in both the upper and lower zones. Unfortunately, a fairly large percentage of patients with pulmonary venous hypertension (PVH) will show a similar blood flow pattern on supine AP chest radiographs because the hydrostatic pressure gradient that develops in the upright position

and produces the typical findings of cephalization of pulmonary blood flow in PVH is lost in the supine position. As a result, it is frequently not possible to distinguish between PVH and overcirculation of the pulmonary vascularity on supine AP chest radiographs.

B. Radiographic signs of overcirculation of the pulmonary vascularity in infants

1. The pulmonary arteries and veins are increased in size in both the upper and lower lung zones with a "balanced" overcirculation pattern with a 1:1 blood flow ratio.
2. The descending branch of the right pulmonary artery is larger than the transverse diameter of the trachea. This is often the best sign of overcirculation in children.
3. The pulmonary vessels seen through the liver are large and prominent end-on vessels are often seen. This is often the best sign of overcirculation in infants.

IV. Plasma volume overload

A. Causes
1. Iatrogenic fluid overload
2. Renal failure

B. Radiographic signs of plasma volume overload
1. Both conditions have increased pulmonary blood flow and blood volume plus increased systemic blood volume.
2. Overcirculation of the pulmonary vascularity
3. Widened vascular pedicle with enlarged SVC, azygos vein, and RA
4. Main pulmonary artery (MPA) usually normal
5. Both interstitial and alveolar edema, when present, tend to be central and perihilar in distribution and spare the costophrenic angles.
6. Pleural effusion common

V. High-flow syndromes

A. Pregnancy
1. Has increased pulmonary blood flow and blood volume plus increased systemic blood volume.
2. Resembles renal failure and iatrogenic fluid overload
3. Overcirculation of pulmonary vascularity
4. Vascular pedicle may be widened

B. Anemia
1. Has increased pulmonary blood flow and blood volume only
2. Overcirculation of pulmonary vascularity
3. Vascular pedicle normal

C. Thyrotoxicosis
1. Has increased pulmonary blood flow and blood volume only
2. Overcirculation of pulmonary vascularity
3. Vascular pedicle normal

D. Peripheral arteriovenous fistula
1. Has increased pulmonary blood flow and blood volume only
2. Overcirculation of pulmonary vascularity
3. Vascular pedicle normal

VI. Acyanotic L-R shunts
A. Atrial septal defect (ASD)
1. Accounts for about 8% of all cases of congenital heart disease (CHD)
2. ASD is more common in females. M:F = 1:2
3. There are six different types of ASD.
   a. Sinus venosus type
      1) 6% of cases of ASD
      2) Almost always associated with partially anomalous pulmonary venous return from the right upper lobe
      3) The right superior pulmonary vein does not cross the descending branch of the right pulmonary artery to enter the left atrium (LA) as it normally does. Instead, the right superior pulmonary vein enters either the SVC or RA.
   b. Ostium secundum (fossa ovalis) type
      1) Most common form (70%)
      2) Rarely presents in infants
      3) Typically presents in older females
   c. Ostium primum type
      1) 20% of cases
      2) Associated with atrioventricular (AV) canal
   d. Defect at junction of inferior vena cava (IVC) and RA — rare
   e. Defect posterior to fossa ovalis — rare
   f. Defect in region of coronary sinus
      1) Associated with absence of the coronary sinus and entrance of a persistent left SVC directly into the LA
      2) Rare
4. ASD — radiographic findings
   a. Overcirculation of the pulmonary vascularity
      1) In general, a L-R shunt must supply at least 50% of the total pulmonary blood flow (2:1 shunt) before a noticeable increase in pulmonary vascularity occurs on the chest radiograph.
      2) Small ASDs (<50%; <2:1) usually have normal vascularity and no abnormality of the cardiac silhouette or MPA.
      3) Moderate-to-large ASDs (50% or more; 2:1 or more) usually have increased pulmonary vascularity and characteristic changes in the cardiac silhouette and MPA. Most ASDs are large and the shunt may be as high as 3:1.
   b. Normal or small vascular pedicle
   c. Enlarged MPA
   d. Right atrial enlargement (RAE) and right ventricular enlargement (RVE)

    e. There is no left atrial enlargement (LAE) because the LA is decompressed by the ASD. The LV is also normal.

    f. Small aorta

5. ASD is a low-pressure shunt.

    a. In ASD the shunt is occurring at the atrial level and is, therefore, a low-pressure shunt.

    b. L-R shunting in ASD is controlled by two factors.

       1) Low pulmonary vascular resistance — pulmonary vascular resistance is much lower than systemic arterial resistance

       2) Right ventricular (RV) distensibility or compliance — the RV is relatively thin-walled and easily distended and can accommodate increased blood flow and volume

       3) These two factors allow L-R flow to occur.

    c. Although the volume of blood flowing through the lungs may be considerably increased, patients with ASD, even those with large ASDs, are often asymptomatic.

    d. Eisenmenger's syndrome

       1) The development of pulmonary arterial hypertension (PAH) secondary to a L-R shunt is known as Eisenmenger's syndrome.

       2) If untreated, pulmonary artery pressure will eventually become high enough to reverse the shunt resulting in a right-to-left (R-L) shunt. This is usually associated with the onset of cyanosis.

       3) Because of the low-pressure nature of the L-R shunt in ASD, Eisenmenger's syndrome usually does not start to develop until the third decade of life in patients with ASD.

    e. Because of the late onset of clinical symptoms, patients with ASD often do not present until adulthood. As a result, ASD is the most common acyanotic L-R shunt presenting in adulthood. Therefore, if you are presented a case of a L-R shunt in an adult, the most likely diagnosis is ASD.

6. Treatment

    a. Direct surgical closure of the ASD with pericardial patch graft

7. Holt-Oram syndrome (cardiac—upper limb syndrome)

    a. Autosomal dominant transmission of cardiac and skeletal anomalies

    b. ASD most common cardiac defect (66%) — ventricular septal defects and patent ductus arteriosus also reported

    c. Thumbs most commonly affected digit

       1) The thumbs may be short, hypoplastic or absent

       2) The thumbs may have 3 phalanges (triphalangeal thumb)

          a) Distal phalanx in same plane as distal phalanges of other digits

          b) Thumb resembles other fingers

          c) Apposition of thumb and digits difficult if not impossible

          d) The thenar eminence may be hypoplastic

    d. The radius may be hypoplastic or joined to the ulna.

e. Other bony abnormalities include phocomelia, hypoplasia of the bones of the shoulder and pectoral musculature, hemivertebra, spina bifida, scoliosis, and Sprengel's deformity.

B. Ventricular septal defect (VSD)

1. Most common type of CHD (25-30%)
2. M:F = 1:1
3. There are five different types.
   a. Membranous (perimembranous, infracristal) defect
      1) Most common type (80%)
   b. Muscular (trabecular) defect
      1) Second most common type (10%)
      2) May be multiple
   c. Posterior (inlet) defect
      1) Approximately 5-10% of cases
      2) Associated with AV canal
   d. Supracristal (infundibular, conal, outlet) VSD
      1) About 5% of cases
      2) May be associated with aortic valvular regurgitation from prolapse of the right, posterior, or both aortic cusps into the VSD
   e. Cristal defect
4. Associated cardiac abnormalities occur in some cases.
   a. Coarctation of the aorta
   b. ASD and patent ductus arteriosus
   c. Intracardiac obstructing lesions including subpulmonic stenosis, subaortic stenosis, mitral stenosis, and anomalous muscle bundle of the RV
   d. Aortic valvular regurgitation
   e. Incompetent AV valves
   f. Slight increased incidence of right aortic arch
5. VSD — radiographic findings
   a. Overcirculation of the pulmonary vascularity
      1) Small VSDs (<50%; <2:1) usually have normal vascularity and no abnormality of the cardiac silhouette or MPA. A tiny VSD with an inordinately loud murmur is known as the *maladie de Roger*. It has no radiographic findings.
      2) Moderate-to-large VSDs (50% or more; 2:1 or more) usually have increased pulmonary vascularity and characteristic changes in the cardiac silhouette and MPA.
   b. Normal or small vascular pedicle
   c. Enlarged MPA
   d. LAE and right ventricular hypertrophy (RVH) and/or RVE. The RA is not enlarged unless there is superimposed congestive heart failure.
   e. Small aorta

6. VSD is a high-pressure shunt.
   a. In VSD the shunt is occurring at the ventricular level and is a high-pressure shunt. The lungs are subjected to increased flow at significantly increased pressure.
   b. Small and moderate-sized VSDs are associated with a pressure gradient between the LV and RV with LV systolic pressure being higher than RV systolic pressure. The degree of shunting is determined by the size of the VSD and the pressure difference across the defect.
   c. Large VSDs are at least 75% of the size of the aorta. There is no pressure gradient between the LV and RV; LV and RV systolic pressures are equal. The magnitude of the shunt across large VSDs is primarily related to the difference between systemic and pulmonary vascular resistance.

7. Clinical course
   a. Small or moderate-sized VSDs may cause no symptoms initially.
   b. Large VSDs often cause volume overload of the LV resulting in the development of congestive heart failure in the first three months of life.
   c. Eisenmenger's syndrome
      1) Was first described with VSD
      2) Because of the high-pressure nature of the shunt Eisenmenger's syndrome develops much earlier in VSD than it does in ASD.
      3) If untreated, pulmonary artery pressure will eventually become high enough to reverse the shunt resulting in a right-to-left (R-L) shunt.
         a) This is usually associated with the onset of cyanosis.
         b) LAE disappears and the cardiac silhouette decreases in size.
         c) RVE becomes much more pronounced.
         d) The MPA becomes larger, the central pulmonary arteries become larger, and the peripheral pulmonary arteries become constricted giving a "pruned" appearance.
   d. Because of its frequency and the high-pressure nature of the shunt, VSD is the most common acyanotic L-R shunt presenting in childhood or infancy. Therefore, if you are presented a case of a L-R shunt in a child, the most likely diagnosis is VSD.

8. Gasul phenomenon
   a. In some untreated VSDs the patient may develop stenosis of the infundibulum of the right ventricular outflow tract (RVOT) with persistence of the VSD.
   b. This has also been referred to as acquired tetralogy of Fallot.

9. Treatment
   a. The majority of VSDs are small and do not present a serious clinical problem. Spontaneous closure of small VSDs occurs in 24% of cases by 18 months, 50% by 4 years, and 75% by 10 years. Up to 45% of uncomplicated membranous or muscular VSDs close by 14 months of

age. Larger VSDs tend to become smaller over time but spontaneous closure is less common, occurring in only about 50% of cases.
   b. Direct surgical closure with pericardial patch grafting of moderate and large VSDs, or small VSDs that do not spontaneously close, is indicated in order to prevent irreversible PAH. Surgery is indicated if the pulmonary arterial systolic pressure is greater than half the systemic arterial systolic pressure, if mean pulmonary arterial pressure exceeds 25 mm Hg, or if the ratio of pulmonary vascular resistance divided by systemic vascular resistance exceeds 0.3:1.
   c. Pulmonary artery banding, which was once a common early treatment for VSD to help prevent the development of PAH, is not done much anymore. Definitive surgical correction of the VSD is the preferred treatment.

C. Patent ductus arteriosus (PDA)
   1. Primary PDA — 9% of cases of CHD
      a. M:F = 1:2
      b. Radiographically resembles VSD with the exception of additional enlargement of the ascending aorta and aortic knob
         1) Overcirculation of the pulmonary vascularity
            a) Small PDAs (<50%; <2:1) usually have normal vascularity and no abnormality of the cardiac silhouette or MPA.
            b) Moderate-to-large PDAs (50% or more; 2:1 or more) usually have increased pulmonary vascularity and characteristic changes in the cardiac silhouette and MPA.
         2) Normal or small vascular pedicle
         3) Enlarged MPA
         4) LAE and RVH or RVE
         5) The LV may be enlarged.
         6) The ascending aorta and aortic knob are often prominent.
            a) Enlargement of the aorta indicates that the L-R shunt is extracardiac (PDA), whereas, a small aorta indicates that the L-R shunt is intracardiac (ASD, VSD, AV canal)
            b) May be noticeable on angiography or MR
            c) Not always noticeable on plain films. Therefore, PDA and VSD often look identical on chest radiographs.
         7) May see dilatation of the patent ductus between the aortic knob and the left pulmonary artery
            a) Ductus "bump" or "bucket handle" sign
            b) May be hard to distinguish from prominent MPA
         8) Compression of the left pulmonary artery by an enlarged ductus may cause diminished blood flow in the left lung.
         9) May have calcification in the wall of the ductus. This can be either circular or tram-track in nature depending on whether it is visualized *en face* or tangentially.
      c. Like VSD, PDA is a high-pressure shunt because the shunt is occurring at the great vessel level. Similar to VSD, clinical symptoms

and Eisenmenger's syndrome develop much earlier in life than in ASD. Still, PDA is not nearly as common as VSD, so if you are presented a case of a L-R shunt in an infant or child the most likely diagnosis is still VSD. In a case which otherwise looks like a VSD on plain films, enlargement of the aortic knob, the presence of a ductus "bump" or "bucket handle" sign, or decreased pulmonary blood flow in the left lung, which suggests compression of the left pulmonary artery by an enlarged ductus, should cause you to also consider the diagnosis of PDA.

   2. Secondary PDA — 20-35% of premature infants
   3. Treatment
      a. Medical management with indomethacin, an antiprostaglandin, is usually the first-line treatment.
      b. Direct surgical closure of the PDA

D. Atrioventricular (AV) canal (endocardial cushion defect)
   1. 3-4% of CHD
   2. M:F = 1:1.3
   3. There are 3 forms.
      a. Partial AV canal
         1) Has cleft in septal leaflet of mitral valve resulting in mitral regurgitation
         2) Ostium primum ASD
         3) Characterized by "gooseneck" deformity of the LV outflow tract on angiography due to downward displacement of the anterior leaflet of the mitral valve that is abnormally attached to the upper portion of the interventricular septum. This abnormal attachment of the anterior leaflet of the mitral valve to the upper portion of the interventricular septum precludes an interventricular communication.
         4) If the ASD is so large that essentially no septal tissue remains, the term common atrium or single atrium is applied.
      b. Transitional AV canal
         1) Cleft septal leaflets of mitral and tricuspid valves with mitral and tricuspid regurgitation
         2) Ostium primum ASD and high membranous VSD
         3) VSD often closed by prolapsing valve leaflets
      c. Complete AV canal
         1) Most common form (60-70%)
         2) Large ostium primum ASD and large VSD involving the upper muscular portion of the ventricular septum
         3) Common AV valve
   4. In all 3 forms the hemodynamic findings are those of a L-R shunt with mitral regurgitation. Partial AV canal tends to present with findings similar to an ASD, while complete AV canal resembles a large VSD.
   5. AV canal — radiographic findings

John H. Woodring, M.D.

a. Mitral regurgitation contributes to the development of cardiomegaly that is out of proportion to the degree of L-R shunting.
b. Overcirculation of the pulmonary vascularity
c. Normal or small vascular pedicle
d. Enlarged MPA
e. RAE and RVE
f. No LAE because LA is decompressed by the ASD
g. Small aorta

6. Associated conditions
a. Strong association with Down's syndrome
1) 25% of Down's syndrome patients have AV canal which accounts for 50% of cardiac defects seen in Down's syndrome
2) 45% of AV canal patients have Down's syndrome
3) Look for hypersegmentation of sternum and the presence of only 11 pairs of ribs. It is thought that the lower ribs help push the endocardial cushions together during embryologic development and when the 12th ribs are absent endocardial cushion defects are more likely to occur.
b. Asplenia and polysplenia
1) Complete AV canal is present in almost 100% of cases of asplenia.
2) Complete AV canal occurs in about 50% of cases of polysplenia.

7. Clinical course
a. Partial AV canal without mitral regurgitation follows a clinical course similar to an ASD. Mitral regurgitation leads to heart failure with resulting symptoms and growth retardation.
b. Complete AV canal rapidly produces congestive heart failure with early development of Eisenmenger's syndrome.

8. Treatment
a. Partial AV canal
1) Early closure of the ASD
2) Plication of the cleft in the septal leaflet of the mitral valve
b. Complete AV canal
1) As with VSD, early pulmonary artery banding is not done much anymore. Early corrective surgery is the preferred treatment.
2) Closure of VSD and ASD with separate pericardial patch grafts
3) Division of the common AV valve with repair of the mitral and tricuspid valves — valve replacement surgery generally can be avoided

E. **Rare causes of a L-R shunt**
1. Aorticopulmonary window
a. Rare abnormality
b. Also called aorticopulmonary septal defect, aorticopulmonary fenestration, or AP window.
c. Results from faulty formation of the truncoconal (spiral) septum that leaves a large communication between the ascending aorta and MPA.
d. Three forms

1) Type I — proximal communication between ascending aorta and MPA above the origin of the coronary arteries
2) Type II — distal communication between the posterior aspect of the ascending aorta and MPA near the origin of the right pulmonary artery
3) Type III — complete defect with communication of the entire ascending aorta and MPA from the level of the valves to the origin of the right pulmonary artery
   a) Must be distinguished from truncus arteriosus
   b) In AP window type III two distinct semilunar valves are present, there is no VSD, and neither great vessel overrides the ventricular septum.
   c) In truncus arteriosus only a single semilunar valve is present, there is a VSD, and the common trunk overrides the ventricular septum.
   e. Usually manifested by cardiomegaly and marked overcirculation of the pulmonary vascularity. The MPA is usually markedly enlarged.
2. Partial anomalous pulmonary venous return (PAPVR)
   a. Occurs in 0.4-0.7% of population
   b. There are two forms of PAPVR.
      1) PAPVR associated with sinus venosus type ASD
         a) There is PAPVR from the right upper lobe.
         b) The right superior pulmonary vein drains to the SVC or RA instead of the LA.
      2) PAPVR associated with hypogenetic lung (scimitar syndrome)
         a) All or part of a hypogenetic lung may be drained by one or more anomalous pulmonary veins.
         b) The anomalous vein usually drains into the IVC below the right hemidiaphragm; less commonly PAPVR drains into the suprahepatic IVC, hepatic veins, portal vein, azygos vein, coronary sinus, or RA.
         c) PAPVR results in a L-R shunt which is usually asymptomatic unless the shunt is 2:1 or greater.
         d) The curvilinear configuration of the anomalous vein as it courses toward the right hemidiaphragm has been likened to a Turkish sword or scimitar. Thus, when hypogenetic lung and PAPVR occur together the condition is referred to as scimitar syndrome.
         e) Scimitar syndrome is almost exclusively a right-sided phenomenon
3. Ruptured aneurysm of sinus of Valsalva
   a. Rare phenomenon
   b. Sinus of Valsalva aneurysm may rupture into a cardiac chamber producing an aorticocameral fistula.
      1) Usually ruptures into RA or RV
      2) Rupture into the LA or LV is uncommon.

        c. Usually presents with signs of PVH rather than overcirculation of the pulmonary vascularity

    4. Left ventricular-right atrial communication

        a. Very rare

        b. Communication between posterior membranous interventricular septum and the RA

        c. Usually mimics an ASD on plain films with overcirculation of the pulmonary vascularity, enlarged MPA, prominent RA and RV and normal-sized LA and LV

    5. Congenital coronary arteriovenous fistula

        a. Abnormal communication between a coronary artery and any cardiac chamber, a pulmonary artery, the coronary sinus, or SVC

        b. Usually associated with overcirculation of the pulmonary vascularity

        c. Can have a prominent bulge in the left heart border resembling enlargement of the left atrial appendage

 F. Oral board advice

    1. Despite the well-documented characteristic findings of each of the four main types of acyanotic L-R shunts, many cases end up looking very similar.

        a. Increased pulmonary vascularity

        b. Enlarged MPA

        c. Cardiomegaly with prominent right side of the heart. You may not be able to tell whether this represents RAE, RVE, or both.

        d. You may not be able to determine if the LA is normal in size or enlarged.

        e. You may not be able to tell if the ascending aorta and aortic arch are enlarged or normal.

    2. How do you handle this situation? It is simply a numbers game! You go with what is statistically most likely.

        a. If the L-R shunt presents in an adult it is most likely an ASD.

        b. If the L-R shunt presents in a child it is most likely a VSD.

        c. If the patient has Down's syndrome or only 11 pairs of ribs the L-R shunt is most likely an AV canal.

VII. Cyanotic admixture lesions (overcirculation with cyanosis)

 A. Oral board advice

    1. Nothing seems to send greater fear into the hearts of oral board candidates than the discussion of cases of congenital heart disease. As a result of this fear, which is often due in part to a lack of an organized approach to the interpretation of chest radiographs of patients with congenital heart disease, most oral board candidates will ask the board examiner whether the patient whose films are being presented is cyanotic or acyanotic.

    2. Most oral board examiners will gladly answer this question, in fact most will freely provide the examinee with the appropriate clinical information

without being asked and say that this is a cyanotic patient or that this is an acyanotic patient with the following symptoms.

3. Unfortunately, some board examiners will refuse to answer such a question and a few will go so far as to say, "You look at the films and you tell me whether this patient is cyanotic or not!" If this happens, how do you handle the situation?

   a. Ask for accessory clinical information
      1) Ask what the patient's hematocrit is
      2) Cyanotic heart disease is associated with an elevated hematocrit.
      3) Asking this question lets the examiner know that you know something about the subject and he or she may open up more. Also, if you find out what the patient's hematocrit is it gives you some idea of whether the patient is cyanotic or acyanotic.

   b. Radiographic findings associated with cyanotic heart disease
      1) The following list of findings has a high association with cyanotic heart disease. Look closely for these.
         a) Decreased pulmonary vascularity
            i) Almost all patients with congenital heart disease associated with decreased pulmonary vascularity are cyanotic.
            ii) See Chapter 7
         b) Mirror-image right aortic arch
            i) There are several different types of right aortic arch, but there are only 2 main types you need to be familiar with.
            ii) Mirror-image right aortic arch has no aortic diverticulum and, therefore, there is no retrotracheal, retroesophageal "mass" effect on the lateral chest radiograph. Mirror-image right aortic arch has a high association with cyanotic heart disease.
            iii) In right aortic arch with an aberrant left subclavian artery, the aberrant left subclavian artery often arises from an aortic diverticulum that is posterior to both the trachea and esophagus. This diverticulum, or the aberrant left subclavian artery itself, produces a retrotracheal, retroesophageal "mass" effect on the lateral chest radiograph just above the top of the aortic arch that indents the trachea and esophagus posteriorly and often displaces the trachea and esophagus anteriorly. The vast majority of patients with aberrant left subclavian artery type right aortic arch have no associated cardiac defect. There is a slight increase in incidence of acyanotic defects.
            iv) If the board examiner shows you a PA chest radiograph of a patient with a right aortic arch, ask for the lateral view. Then you can tell whether the patient is cyanotic or not.
            v) See Chapter 11
         c) "Figure-of-8" sign or "snowman" sign

   i) This is characteristic of totally anomalous pulmonary venous return (TAPVR) with supracardiac connection to a persistent left superior vena cava.

   ii) These patients are almost always cyanotic.

  d) A deep concavity at the MPA segment

   i) This indicates that the MPA is either in the wrong location (transposition of the great vessels) or is absent (truncus arteriosus types II, III, and IV), that there is significant pulmonic stenosis or atresia (tetralogy of Fallot, pulmonic stenosis or atresia with an intact ventricular septum), or that there is significant obstruction to flow through the right side of the heart (Ebstein's anomaly, tricuspid atresia).

   ii) All of these conditions are associated with cyanosis.

   iii) If the patient has decreased vascularity with a concavity at the MPA segment the patient either has tetralogy of Fallot, pulmonic stenosis or atresia with an intact ventricular septum, Ebstein's anomaly, or tricuspid atresia. These can be separated from each other based upon whether the heart size is normal (tetralogy of Fallot) or is enlarged, particularly if there is enlargement of the RA (pulmonic stenosis or atresia with an intact ventricular septum, Ebstein's anomaly, or tricuspid atresia). See Chapter 7. If the patient has overcirculation of the pulmonary vascularity with a concavity at the MPA segment the patient probably has transposition of the great vessels or truncus arteriosus types II, III, or IV. Some cases of single ventricle with severe pulmonic stenosis also have this appearance. See later sections of this chapter.

  e) Dextrocardia

   i) Is often associated with cyanotic heart disease, particularly in patients with abnormal situs.

   ii) See Chapter 9

  f) PAH

   i) Whether it is due to Eisenmenger's syndrome or other conditions, PAH is often associated with cyanosis.

   ii) See Chapter 6

2) Whether you think it is reasonable or not, you are expected to know the above information. If you do, you can handle almost any case of congenital heart disease shown to you on oral boards. In fact, any board examiner who asks you to look at the film and tell him or her whether the patient is cyanotic or not has to be showing you one of the above conditions! If they aren't they can't ask you that question.

c. Some cases of cyanotic and acyanotic heart disease look almost identical. For example, supracardiac TAPVR without a characteristic "snowman" or 'figure-of-8" looks almost identical to an ASD (increased vascularity, enlarged MPA, prominent right heart, and no LAE). In the worst case scenario (the board examiner won't tell you whether the patient is cyanotic or acyanotic and won't tell you the patient's hematocrit, and you can't find anything on the films that would give you a clue as to whether the patient was cyanotic or acyanotic) the best thing to do is to identify as many pertinent findings as possible and then list those you would consider if the patient were cyanotic and those you would consider if the patient were acyanotic. This will at least show the examiner that you have an organized approach to cardiac disease. However, you shouldn't get to this point, because most cases shown on oral boards have characteristic findings if you know what to look for.

VIII. **Differential diagnosis of cyanotic admixture lesions (cyanosis with overcirculation of the pulmonary vascularity) —**
mnemonic = 3 "T's" and an "S"
  A. Transposition of the great vessels (TGV)
    1. 3-4% of cases of CHD
    2. Males are more commonly involved than females. M:F = 2:1 to 3:1
    3. Cyanosis develops in the neonatal period and is often severe. TGV is the most common cause of cyanosis in a neonate.
    4. The great vessels originate from the inappropriate ventricle. The aorta arises from the RV and the MPA arises from the LV.
    5. There are 2 forms.
      a. Complete TGV (D-transposition, D-TGV)
        1) There is ventriculoarterial discordance in the presence of atrioventricular concordance. In other words, the atria and ventricles are properly connected; the aorta and the MPA arise from the wrong ventricle.
        2) Systemic venous blood returns to the anatomic RA, flows through the tricuspid valve into the RV, and then flows out of the RV via the aorta into the systemic arterial circulation. Pulmonary venous blood returns to the anatomic LA, flows through the mitral valve into the LV, and then flows out of the LV via the MPA to the pulmonary arterial circulation.
        3) As can be seen, the systemic and pulmonary circulations are parallel and separate — this situation requires mixing of blood between the two circulations (admixture lesion) to support life.
          a) May have an intact ventricular septum (50% of cases)
            i) Admixture occurs through a PDA and a patent foramen ovale.
            ii) When the ductus closes at about 48 hours of life this leaves only a small patent foramen ovale for bi-directional shunting to occur. This is usually inadequate to support life and the patient will die in a few days unless some type of

communication is created to allow intracardiac mixing to occur.

        iii) Patients usually have profound cyanosis.

    b) VSD (50% of cases)

        i) Intracardiac mixing of blood is better and patients are usually less cyanotic.

        ii) In those without pulmonic stenosis, congestive heart failure tends to occur early as in isolated large VSDs. Eisenmenger's syndrome also develops early.

        iii) In those with pulmonic stenosis, the pulmonic stenosis limits the severity of the pulmonary overcirculation and helps prevent congestive heart failure. Pulmonic stenosis also favors shunting of oxygenated blood from the LV into the aorta.

4) A mirror-image right aortic arch is present in 3% of cases.

5) Complete TGV — radiographic findings

    a) Overcirculation of the pulmonary vascularity

    b) Cardiomegaly typically involving all four chambers

    c) Concave MPA segment — MPA is posterior and in the midline and does not produce a border with lung

    d) The mediastinum is narrow due to involution of the thymus from severe stress. The combination of the enlarged cardiac silhouette and narrow mediastinum produces a characteristic configuration likened to an "egg-on-its-side" or an "apple-on-a-string."

    e) Right aortic arch occasionally seen.

6) Treatment

    a) Complete TGV with intact ventricular septum

        i) Prostaglandin E_1 to maintain patency of the ductus

        ii) Rashkind procedure — balloon septostomy to create ASD

        iii) Definitive repair with arterial switch procedure

        iv) Mustard or Senning procedures (removal of the atrial septum and creation of an intraatrial baffle directing pulmonary venous return to the RV and systemic venous return to the LV) are not done much anymore. Definitive repair with arterial switch procedure is the preferred treatment.

    b) Complete TGV with VSD

        i) Pulmonary artery banding may be done for patients with large VSDs and no pulmonic stenosis to prevent PAH prior to definitive surgical correction.

        ii) Definitive repair with arterial switch procedure and closure of the VSD

b. Congenitally corrected TGV (L-transposition, L-TGV)

    1) Rare

2) There is ventriculoarterial discordance in the presence of atrioventricular discordance resulting in doubly switched circulations. Not only do the aorta and MPA connect to the wrong ventricle but the ventricles are also connected to the wrong atria. The atrioventricular valves follow their corresponding ventricles so that the anatomic mitral valve is on the right and the anatomic tricuspid valve is on the left.

3) Systemic venous blood flows into the anatomic RA, flows through the anatomic mitral valve that is on the right side into the anatomic LV that is also positioned on the right side, and then flows out of the LV via the MPA into the pulmonary circulation. Pulmonary venous blood flows into the anatomic LA, flows through the anatomic tricuspid valve that is on the left side into the anatomic RV that is also positioned on the left side, and then flows out of the RV via the aorta into the systemic circulation.

4) Thus the abnormalities are congenitally "corrected" (two wrongs make a right). Unfortunately, the heart is usually far from normal.

5) Rarely, there may be no other associated cardiac abnormality, and if this is the case, the pulmonary vascularity will be normal. However, associated cardiac abnormalities are common and include the following.

a) Insufficiency of the left-sided atrioventricular (anatomic tricuspid valve) resulting in clinical features of "mitral" regurgitation.

b) VSD (75% of cases)
   i) Usually perimembranous in location
   ii) Often large
   iii) Can lead to Eisenmenger's physiology

c) Pulmonic stenosis or atresia (40%)
   i) Usually associated with VSD
   ii) Usually subvalvular, rarely valvular

d) Complete heart block (33%)

e) The RV is not anatomically designed to be subjected to systemic vascular resistance and systemic arterial pressure for life. Even those patients who have no associated congenital cardiac abnormality develop congestive heart failure much earlier in life than the normal population.

6) Corrected TGV — radiographic findings

a) The pulmonary vascular pattern is determined by the associated cardiac abnormalities. If there are no associated cardiac abnormalities the pulmonary vascularity will be normal. Most cases demonstrate overcirculation of the pulmonary vascularity because of an associated VSD. If there is severe pulmonic stenosis or atresia the vascularity will be decreased.

b) The cardiac configuration is characteristic. The RVOT, which is left-sided, and the ascending aorta arising from it, produce a broad convexity along the left heart border that extends from the cardiac base to the aortic knob.

    c) The MPA segment is usually not visible.

  7) Treatment

    a) Conventional approach

      i) Closure of VSD in those with isolated VSD and closure of VSD and formation of a conduit from the left (pulmonary) ventricle to the pulmonary artery in those with VSD and associated pulmonic stenosis or atresia.

      ii) This has not met with good results in many cases because of complete heart block, left AV valve regurgitation, and long-term right (systemic) ventricular dysfunction leading to congestive heart failure.

    b) Newer techniques

      i) In those with isolated VSD an arterial switch procedure is performed. Since this results in surgical creation of complete TGV, an atrial switch, such as a Mustard or Senning procedure, is also performed.

      ii) In those with VSD and pulmonic stenosis or atresia, the VSD is closed in a way that diverts blood from the LV into the aorta and from the RV via a conduit into the pulmonary artery. This also creates complete transposition physiology, so an atrial switch procedure is done as well.

B. Truncus arteriosus

  1. About 2% of cases of CHD

  2. Results from failure of formation of the truncoconal (spiral) septum within the truncus in the early embryo

  3. As a result, a single arterial vessel (the truncus) leaves the heart.

  4. In 70% of cases the truncal valve is tricuspid, but there can be from 2-5 cusps.

  5. The truncus sits atop a large VSD.

  6. Overcirculation of the pulmonary vascularity is often severe and leads to early development of Eisenmenger's syndrome.

  7. There are said to be four types; actually there are only three.

    a. Type I (60%)

      1) Most common form

      2) A short MPA arises from the left side of the truncus and divides into right and left pulmonary arteries.

      3) Because the MPA is anatomically present, the chest radiograph frequently shows an enlarged MPA.

      4) There is moderate cardiomegaly with enlargement of both ventricles and the LA.

      5) Overcirculation of the pulmonary vascularity

    b. Type II (30%)

      1) Each pulmonary artery arises independently from the posterolateral aspect of the left side of the truncus.

        2) There is no MPA. The MPA segment is concave.

        3) There is moderate cardiomegaly with enlargement of both ventricles and the LA.

        4) Overcirculation of the pulmonary vascularity

  c. Type III (10%)

        1) The right pulmonary artery arises from the right lateral wall of the truncus; the left pulmonary artery arises from the left lateral wall of the truncus.

        2) There is no MPA. The MPA segment is concave.

        3) There is moderate cardiomegaly with enlargement of both ventricles and the LA.

        4) Overcirculation of the pulmonary vascularity

  d. Type IV (pseudotruncus) (see Chapter 8)

        1) Actually represents tetralogy of Fallot with pulmonic valvular atresia.

        2) The pulmonic valve is atretic; however, the right and left pulmonary arteries are present. Pulmonary blood flow comes from large systemic arterial collaterals, usually bronchial arterial collaterals, which arise from the descending thoracic aorta and connect to the right and left pulmonary arteries.

8. Associated anomalies

  a. Mirror-image right aortic arch common (40%)

        1) Differential diagnosis of cyanotic heart disease, and mirror-image right aortic arch

            a) Associated with overcirculation of the pulmonary vascularity

                i) Truncus arteriosus — 40% of cases have mirror-image right aortic arch

                ii) TGV — 3% of cases have mirror-image right aortic arch

                iii) Tricuspid atresia — 5-10% of cases of tricuspid atresia have mirror-image right aortic arch; however, only 20% of cases of tricuspid atresia have overcirculation of the pulmonary vascularity

            b) Associated with decreased pulmonary vascularity

                i) Tetralogy of Fallot — 25% of cases have mirror-image right aortic arch

                ii) Tricuspid atresia — again 5-10% of cases of tricuspid atresia have mirror-image right aortic arch; 80% of cases of tricuspid atresia have decreased pulmonary vascularity.

            c) The highest percentage of cases with mirror-image right aortic arch occurs in truncus arteriosus. However, since tetralogy of Fallot is much more common, most of the cyanotic patients you will encounter in your practice with a mirror-image right aortic arch will have tetralogy of Fallot.

            d) Based just upon the presence of a mirror-image right aortic arch and whether the pulmonary vascularity is increased or decreased, you can make a fairly accurate prediction of what the underlying cardiac abnormality is. In mirror-image right aortic arch associated with overcirculation of the pulmonary vascularity the patient probably has truncus arteriosus or possibly TGV. If the pulmonary vascularity is decreased the patient probably has tetralogy of Fallot or possibly tricuspid atresia.

    b. Aortic arch interruption (20%)
       1) Associated with a large PDA
       2) The PDA supplies the left subclavian artery, most of the thorax, and the lower part of the body.
       3) When the PDA closes there is no blood supply to most of the thorax and abdomen.
    c. Pulmonic stenosis (14%)
       1) Protects patients from developing severe overcirculation and Eisenmenger's syndrome
       2) Pressure in the pulmonary artery is less than that in the truncus.
  9. Treatment
    a. Closure of the VSD leaving the truncus arising from the LV
    b. The pulmonary arteries are detached from the truncus and connected to the RV using a valved conduit.

C. Totally anomalous pulmonary venous return (TAPVR)
  1. 1% of cases of CHD
  2. ASD is an integral component of the anomaly.
  3. The four pulmonary veins do not converge on the LA; instead, they converge on a "confluence" of pulmonary veins, known as the common pulmonary vein, which maintains a connection to the primitive systemic venous system.
  4. There are four types.
    a. Supracardiac connection — 55%
       1) Connection to a right superior vena cava (RSVC) (10%)
          a) Represents persistence of communication to the right anterior cardinal system (supracardiac connection)
          b) There is overcirculation of the pulmonary vascularity.
          c) The MPA is often enlarged.
          d) The upper right margin of the mediastinum is prominent and convex outwards due to enlargement of the SVC that is carrying both systemic venous return from the upper body and pulmonary venous return from both lungs.
          e) The cardiac silhouette is enlarged with both RAE and RVE.
          f) The ASD allows R-L shunting of admixed blood to the LA; however, the LA is not enlarged.
       2) Connection to a left superior vena cava (LSVC) or "vertical vein" (45%)
          a) Represents persistence of communication to the distal part of the left anterior cardinal system (supracardiac connection)
          b) There is overcirculation of the pulmonary vascularity.

      c) There is a characteristic "snowman" or "figure-of-8" configuration of the heart and upper mediastinum due to a convex outward bulge in the upper left mediastinal border from the prominent LSVC or vertical vein and a convex outward bulge in the upper right mediastinal border from the enlarged RSVC. Blood flows from the common pulmonary vein up the LSVC or vertical vein, crosses the mediastinum via the left brachiocephalic or innominate vein, and enters the RSVC that is also enlarged. These enlarged veins produce the upper half of the "snowman" and the heart forms the lower half. These enlarged veins also produce a pretracheal soft-tissue "mass effect" on the lateral chest radiograph.

      d) The MPA is enlarged but this may be obscured on the plain film by the upper half of the "snowman."

      e) The cardiac silhouette is enlarged with both RAE and RVE.

      f) The ASD allows R-L shunting of admixed blood to the LA; however, the LA is not enlarged.

  b. Cardiac connection — 30%

    1) Represents persistence of communication to the proximal part of the left anterior cardinal system (cardiac connection)

    2) Connection to the coronary sinus (16%)

    3) Connection to RA (14%)

    4) There is no characteristic cardiac configuration. The radiographic appearance is identical to ASD.

      a) Overcirculation of pulmonary vascularity

      b) Enlarged MPA

      c) Cardiomegaly with RAE, RVE, and no LAE

  c. Connection to the portal venous system with obstruction — 12%

    1) Represents persistence of communication to the umbilical-vitelline system (infracardiac connection)

    2) The anomalous right and left pulmonary veins converge into a single venous trunk that follows the esophagus through the esophageal hiatus in the diaphragm into the abdomen to connect to the portal venous system. Pulmonary venous return is typically obstructed at the level of the diaphragm where the anomalous pulmonary vein courses through the esophageal hiatus.

    3) Characterized by severe reticulation

    4) The heart size is normal. The MPA is not prominent.

  d. Mixed lesions — 3%

5. Associated defects

  a. PDA

  b. VSD

  c. Coarctation of the aorta

  d. Situs ambiguus (asplenia and polysplenia)

    1) AV canal

    2) Single ventricle

6. Treatment
   a. Creation of large communication between the LA and pulmonary venous system
   b. Obliteration of the anomalous pulmonary venous connection to the systemic circulation
   c. Closure of the ASD

D. <u>Single ventricle</u>
1. 0.3% of cases of CHD
2. Also know as univentricular heart. A single ventricle is a heart with two atria but only one ventricle.
3. There are several types of single ventricle depending on whether the LV or the RV is the predominant ventricle.
4. The predominant chamber is the inlet chamber; the rudimentary chamber is the outlet chamber.
5. The AV valves (tricuspid and mitral valves) have 5 ways to connect with the single ventricle.
   a. Two patent AV valves connecting with the single ventricle
      1) Double inlet LV
      2) Double inlet RV
   b. Right AV valve atresia
   c. Left AV valve atresia
   d. Common AV valve (common in asplenia)
   e. Straddling of the right or left AV valve into the rudimentary ventricular chamber
6. LV predominance (single ventricle of LV type) — most common form (70%)
   a. The main ventricular chamber has the trabecular pattern of a LV and communicates with the rudimentary RV via a foramen, the bulboventricular foramen, which is analogous to a VSD.
   b. The rudimentary RV can be found at the base of the heart on the right side (normally related infundibular chamber) or the left side (inverted infundibular chamber). Inverted rudimentary RV is most common.
   c. The rudimentary RV may give rise to the both great vessels, one great artery, or occasionally no great artery.
   d. Most often the ascending aorta arises from the rudimentary RV and the MPA arises from the LV. Thus the great arteries are almost always transposed (90% of cases).
   e. When the outlet chamber is on the right, the aorta is in the position of D-TGV. When the outlet chamber is on the left, the aorta is on the position of L-TGV.
      1) Single ventricle with D-TGV and straddling of the right-sided AV valve into the rudimentary chamber is called "Lambert heart."
      2) The rare configuration of a single ventricle with the MPA arising from the outlet chamber (normally related great vessels) is called "Holmes heart."

3) If no great vessel arises from the rudimentary ventricle the chamber is called a ventricular pouch.
  f. The pulmonic valve may be normal, or there may be pulmonic atresia or pulmonic stenosis.
  g. Radiographic findings
    1) Cardiomegaly with overcirculation of the pulmonary vascularity is present in most cases. Only those with severe pulmonic stenosis or atresia have a normal size heart and decreased vascularity.
    2) When the rudimentary RV is inverted, the rudimentary RV produces a characteristic small bulge along the upper left heart border usually where the left atrial appendage would normally be located.
    3) "Waterfall" sign of right hilum
      a) Can occur in single ventricle
      b) Is due to a large jet of blood that flows directly from the LV into the right pulmonary artery and grossly enlarges the right pulmonary artery and its segmental branches.
      c) Other causes of a "waterfall" right hilum
        i) Truncus arteriosus
        ii) TGV
7. RV predominance with mitral stenosis or atresia (single ventricle of RV type)
  a. 20% of cases
  b. The main ventricular chamber has the trabecular pattern of an RV
  c. Part of hypoplastic left heart syndrome
  d. See Chapter 4
8. If the main chamber resembles neither an RV nor LV, an undifferentiated or primitive single ventricle is present (10%).
9. If characteristics resembling both ventricles are present, but there is no interventricular septum or only a tiny hypoplastic interventricular septum, the condition is called a common ventricle. This is felt to represent an extreme manifestation of a VSD rather than a true single ventricle.
10. Associated defects
  a. Pulmonic stenosis
  b. Aortic stenosis
  c. D-TGV
  d. L-TGV
  e. Coarctation of the aorta
  f. Interruption of the aorta
  g. Asplenia
  h. Hydroureter
  i. Imperforate anus
11. Treatment
  a. For those with pulmonic stenosis or atresia a Blalock-Taussig shunt (side-to-side anastomosis of the subclavian artery and pulmonary artery on the side opposite the aortic arch) can be done initially.

   b. Glenn procedure (anastomosis of SVC to right pulmonary artery) at 6-18 months of age
   c. Modified Fontan procedure (conduit between the right atrium and pulmonary artery) at 1-3 years of age

E. Others
   1. The following conditions can also be associated with cyanosis and overcirculation of the pulmonary vascularity.
   2. Tricuspid atresia
      a. 1.2% of cases of CHD
      b. Typically has decreased vascularity
      c. If there is associated TGV or large VSD, with no pulmonary stenosis, there is overcirculation of the pulmonary vascularity (seen in about 20% of cases of tricuspid atresia).
   3. Common atrium
      a. Rare
      b. The atrial septum is virtually completely absent.
      c. There may be clefts in the mitral and tricuspid valves.
      d. There is a bi-directional shunt (admixture) with cyanosis.
   4. Double-outlet right ventricle (DORV)
      a. 0.8% of cases of CHD
      b. In DORV both the aorta and MPA arise from the RV. A VSD is the only outlet from the LV.
      c. The aortic valve is almost always to the right of the pulmonic valve and the aortic and pulmonic valves are usually side-by-side.
      d. Taussig-Bing malformation refers to DORV with a subpulmonic VSD, with or without overriding of the pulmonary valve above the ventricular septum, without pulmonic stenosis. Accounts for 32% of cases of DORV.
      e. Can have subaortic VSD
         1) With pulmonic stenosis — 35%
         2) Without pulmonic stenosis — 23%
      f. In 10% the VSD is far from both great vessels (remote, indeterminate, or noncommitted VSD).
      g. DORV — radiographic findings
         1) The findings are not characteristic.
         2) If there is no pulmonic stenosis, there will be cardiomegaly with RVE and overcirculation of the pulmonary vascularity. The MPA segment may be unapparent but is often enlarged and the radiographic appearance resembles that of a VSD.
         3) If there is pulmonic stenosis, there will be cardiomegaly with RVE and normal or decreased pulmonary blood flow. The radiographic picture thus mimics tetralogy of Fallot.
         4) There may be a mirror-image right aortic arch.
      h. Associated cardiac defects
         1) Pulmonic stenosis, both valvular and infundibular
         2) Valvular or subvalvular aortic stenosis

   3) Coarctation of the aorta
   4) Interruption of the aorta
   5) AV canal
   6) Mitral stenosis or atresia
   7) Parachute mitral valve
   8) Supravalvular mitral ring
   9) Straddling of the left or right AV valves
   10) AV septal malalignment
   i. DORV may also be part of the situs ambiguus syndrome. In these patients DORV may be associated with endocardial cushion defects, single ventricle, and abnormalities of systemic and pulmonary venous return.
5. Double-outlet left ventricle (DOLV)
   a. DOLV is defined as an abnormal ventriculoarterial alignment in which both great arteries arise above the LV.
   b. Very rare anomaly
   c. In the most common form, there is situs solitus with normal ventricles, D-TGV, subaortic VSD, and pulmonic stenosis.
   d. Can also have subpulmonic VSD, situs inversus, situs ambiguus (asplenia and polysplenia), atrioventricular discordance, tricuspid atresia, mitral atresia, straddling tricuspid valve, Ebstein's anomaly, single ventricle, aortic stenosis, hypoplastic ascending aorta, and preductal coarctation of the aorta.

References and suggested additional reading

1. Coussement AM, Gooding CA. Objective radiographic assessment of pulmonary vascularity in children. Radiology 1973; 109:649-654.
2. Dähnert W. *Radiology Review Manual 4th ed*. Philadelphia: Lippincott Williams & Wilkins, 2000.
3. Elliott LP. Roentgenologic approach to heart disease: Stage I — extracardiac analysis. In: Taveras JM, Ferrucci JT eds. *Radiology: Diagnosis-Imaging-Intervention*. Volume 2, Chapter 10. Hagerstown: Lippincott, 1991.
4. Elliott LP. Stage II — Physiologic stage of analysis. In: Taveras JM, Ferrucci JT eds. *Radiology: Diagnosis-Imaging-Intervention*. Volume 2, Chapter 11. Hagerstown: Lippincott, 1991.
5. Elliott LP. Pathophysiology and roentgenologic findings in persistent common atrioventricular canal. In: Taveras JM, Ferrucci JT eds. *Radiology: Diagnosis-Imaging-Intervention*. Volume 2, Chapter 87. Hagerstown: Lippincott, 1991.
6. Fuster V, Alexander RW, O'Rourke RA, Roberts R, King SB III, Wellens HJJ. *Hurst's The Heart 10th ed*. New York: McGraw-Hill, 2001.
7. Gedgaudas E, Moller JH, Castaneda-Zuniga WR, Amplatz K. *Cardiovascular Radiology*. Philadelphia: Saunders, 1985.
8. Meszaros WT. *Cardiac Roentgenology: Plain Films and Angiocardiographic Findings*. Springfield: Charles C Thomas, 1969.

John H. Woodring, M.D.

9. Miller SW. *Cardiac Radiology: The Requisites*. St. Louis: Mosby, 1996.
10. Milne ENC, Pistolesi M. *Reading the Chest Radiograph: a Physiologic Approach*. St. Louis: Mosby, 1993.
11. Milne ENC, Pistolesi M, Miniati M, Giutini C. The vascular pedicle of the heart and the vena azygos. Part 1: the normal subject. Radiology 1984; 152:1-8.
12. Pistolesi M, Milne ENC, Miniati M, Giutini C. The vascular pedicle of the heart and the vena azygos. Part 2: acquired heart disease. Radiology 1984; 152:9-17.
13. Spindola-Franco H, Fish BG. *Radiology of the Heart: Cardiac Imaging in Infants, Children, and Adults*. New York: Springer-Verlag, 1985.
14. Woodring JH. Pulmonary artery-bronchus ratios in patients with normal lungs, pulmonary vascular plethora, and congestive heart failure. Radiology 1991; 179:115-122.

Chapter 6.

Pulmonary arterial hypertension

I. **Pathophysiology of pulmonary arterial hypertension (PAH)**
A. **Laplace's law**
1. When PAH develops there is a pathognomonic structural alteration in the pulmonary arteries.
a. The central pulmonary arteries, characteristically the main pulmonary artery (MPA), central right and left pulmonary arteries, and the descending branches of the right and left pulmonary arteries, become dilated.
b. In contrast to this, the lobar arteries and those arteries beyond the lobar division decrease in caliber.
c. In general, the degree of central distention reflects the severity of PAH.
2. Laplace's law offers the best explanation for the paradoxical central pulmonary arterial dilatation and peripheral vasoconstriction that characterizes PAH. Laplace's law, which is explained by the following formula:

$$T = P \times r$$

where T is the tension on the vessel wall, P is the intraluminal pressure, and r is the radius of the vessel, states that the tension on the wall of a vessel is the product of the intraluminal pressure and the radius of the vessel. As vessel size increases, wall tension increases for any given intraluminal pressure. As intraluminal pressure increases, greater wall tension is placed on the larger vessels than on the smaller vessels. In response to an initial increase in intraluminal pressure, vessels smaller than 1 cm in diameter are still exposed to a relatively low wall tension and can respond by vasoconstriction. Proximal pulmonary arteries larger than 1 cm cannot overcome the increased wall tension by vasoconstriction, and they passively dilate instead. Typically, this transition occurs at the level of the lobar arteries.
3. If the lobar arteries are enlarged to begin with, and are larger than the critical 1 cm diameter, they too will fail to constrict and the transition point will move out to the segmental level. This situation frequently occurs in patients with preexisting left-to-right (L-R) intracardiac shunts.
B. **Diagnostic criteria for PAH**
1. Pulmonary arterial systolic pressure greater than 30 mm Hg
2. Mean pulmonary arterial pressure greater than 18 mm Hg

C. Pathology of PAH

1. Grade I — medial hypertrophy
2. Grade II — concentric or eccentric cellular intimal proliferation
3. Grade III — relatively acellular intimal fibrosis with occlusion of the smaller pulmonary arteries and arterioles
4. Grade IV — progressive, generalized dilatation of the distal muscular arteries and the appearance of plexiform lesions (complex vascular structures composed of a network or plexus of proliferating endothelial tissue, frequently accompanied by thrombus, within a dilated thin-walled sac)
5. Grade V — thinning and fibrosis of the media superimposed on the plexiform lesions
6. Grade VI — necrotizing arteritis within the media
7. Grades I and II may return to normal if the cause of the PAH can be corrected (such as a L-R shunt); however, grades III-VI usually do not. A pulmonary vascular resistance to systemic vascular resistance ratio (R_p/R_s) of 0.7:1 is generally considered a contraindication to surgery.

II. Causes of precapillary PAH

A. Persistent fetal circulation

1. Occurs in infants at or near term
2. Characterized by tachypnea and cyanosis associated with abnormal postnatal persistence of PAH and a right-to-left shunt through the foramen ovale, the ductus arteriosus, or both.
3. Usually associated with right ventricular failure
4. Predisposing factors — chronic fetal distress resulting in hypertrophy of the pulmonary arterial musculature
   a. Gestational diabetes
   b. Perinatal asphyxia
   c. Aspiration
   d. Hyperviscosity
   e. Hypocalcemia
   f. Hypoglycemia
   g. Acidosis
5. Radiographic findings
   a. Cardiomegaly with right atrial and right ventricular enlargement
   b. Marked enlargement of the MPA and central pulmonary arteries with peripheral pruning
6. Some affected infants die from progressive hypoxemia, others survive with complete resolution of the PAH.

B. Eisenmenger's syndrome

1. Acyanotic L-R shunts
   a. Atrial septal defect
   b. Ventricular septal defect
   c. Patent ductus arteriosus
   d. Atrioventricular canal

     e. Aorticopulmonary window
2. Cyanotic admixture lesions associated with overcirculation of the pulmonary vascularity
     a. Transposition of the great vessels
     b. Truncus arteriosus
     c. Totally anomalous pulmonary venous return
     d. Single ventricle
     e. Others
          1) Tricuspid atresia with transposition of the great vessels and no pulmonic stenosis
          2) Common atrium
          3) Double-outlet right ventricle

C. Supravalvular/peripheral pulmonic stenosis
1. Can occur as an isolated anomaly
2. Can be part of William's syndrome
     a. Mental retardation
     b. Hypercalcemia
     c. Elfin facies
     d. Supravalvular/peripheral pulmonic stenosis
     e. Supravalvular aortic stenosis
3. There are four types.
     a. Type I — stenosis of the MPA
     b. Type II — stenosis at the bifurcation of the MPA which causes narrowing of the distal MPA and proximal right and left pulmonary arteries
     c. Type III — multiple peripheral stenoses with normal MPA and normal central right and left pulmonary arteries
     d. Type IV — central and peripheral stenoses often resulting in diffuse hypoplasia of the MPA and central pulmonary arteries with associated peripheral stenoses
4. The chest radiographic appearance is variable depending upon which type of supravalvular pulmonic stenosis the patient has. If the stenosis is severe, the pulmonary vascularity may be decreased, but it frequently appears relatively normal. The MPA can be enlarged but is usually not enlarged and may be unapparent. The central right and left pulmonary arteries and the descending branches of the right and left pulmonary arteries can also be enlarged but are more often normal or decreased in size. Occasionally the pulmonary arteries have a "beaded" or "sausage-like" appearance from areas of stenosis with distal post-stenotic dilatation.
5. The diagnosis is usually made by angiography although MR can also confirm the diagnosis.

D. Idiopathic (primary PAH)
1. Rare — incidence is 1/1,000,000
2. Slightly more common in women; M:F = 1:1.7
3. Average age at presentation is 36 years.

E. Pulmonary embolism
1. Bland pulmonary thromboembolism
2. Tumor embolism
3. Schistosomiasis
   a. Can cause massive occlusion of the pulmonary arterial tree from embolization of migrating parasites
   b. Can result in massive cardiomegaly

F. Hepatic disease
1. PAH is seen occasionally following portosystemic shunting in patients who have non-cirrhotic portal hypertension secondary to portal fibrosis or multifocal nodular hyperplasia.

G. Systemic immunologic disorders and vasculitis
1. Systemic lupus erythematosus
2. Mixed connective tissue disorder
3. Scleroderma
4. Rheumatoid lung
5. Pulmonary vasculitis
   a. Polyarteritis nodosa
   b. Takayasu's arteritis
   c. Behçet's disease

H. Pulmonary capillary hemangiomatosis

I. Primary lung diseases
1. Emphysema
2. Chronic bronchitis
3. Bronchiectasis
4. Cystic fibrosis
5. Bronchiolitis
6. End-stage (honeycomb) lung or severe pulmonary fibrosis of any cause

J. Pleural fibrosis (fibrothorax)

K. Chest wall deformities
1. Severe kyphoscoliosis
2. Thoracoplasty

L. Alveolar hypoventilation
1. Obesity
2. Obstructive sleep apnea
3. Neuromuscular disease
4. Idiopathic (Ondine's curse)

M. High altitude

N. HIV disease

III. Causes of postcapillary PAH
A. Chronic left ventricular failure of any cause
B. Obstruction of the mitral valve
1. Mitral stenosis

2. Left atrial myxoma

C. Obstruction of the pulmonary venous system proximal to the mitral valve
1. Cor triatriatum
2. Congenital stenosis of the pulmonary veins
3. Pulmonary veno-occlusive disease
4. Fibrosing mediastinitis

IV. PAH — radiographic findings
A. Enlarged MPA
1. Seen as a marked convexity of the MPA segment on plain films
2. Diameter of MPA 3 cm or more on CT or MR
B. Enlarged central right and left pulmonary arteries
C. Enlargement of the descending branches of the right and left pulmonary arteries
1. The upper limit of normal for the width of the descending branches of the right and left pulmonary arteries is 1.5 cm in women and 1.6 cm in men.
2. Because of extreme variation in patient size within the normal population, I have found that using values of 1.6 cm in women and 1.7 cm in men, as indicators of an enlarged descending branch of the right or left pulmonary artery, is unreliable and results in an extremely high number of false positive diagnoses of PAH. In general, I have found that using a value of 2 cm or greater, as an indicator of an enlarged descending branch of the right or left pulmonary artery is more reliable. Although some cases of mild PAH might be missed, the 2 cm value does not result in an unacceptable number of false positive diagnoses of PAH.
3. In most causes of PAH the transition between central dilatation and peripheral vasoconstriction occurs at the level of the lobar arteries.
   a. The MPA, central right and left pulmonary arteries, and descending branches of the right and left pulmonary arteries are enlarged.
   b. The lobar arteries, segmental arteries, and the arteries beyond the segmental vessels are constricted.
4. In Eisenmenger's syndrome the transition point extends out to the lobar arteries, since they were enlarged to begin with and could not constrict in response to increased intraluminal pressure.
   a. The MPA, central right and left pulmonary arteries, descending branches of the right and left pulmonary arteries, and lobar arteries are enlarged.
   b. Peripheral vasoconstriction is limited to vessels at the segmental level and beyond.
   c. This difference should allow you to be able to separate PAH secondary to Eisenmenger's syndrome from PAH of other causes on plain films.
D. Constricted or "pruned" peripheral arteries

E. Enlarged right heart
1. Initially there is right ventricular hypertrophy, followed later by right ventricular enlargement.
2. Once right ventricular failure ensues there will be right atrial enlargement and other signs of right ventricular failure which include the following:
   a. Widened vascular pedicle with enlargement of the superior vena cava and azygos vein
   b. Posterior bulge of the suprahepatic portion of the inferior vena cava (IVC) on lateral chest radiograph
   c. On CT scans of the abdomen there will be passive congestion of the liver, enlargement of the IVC and hepatic veins, and ascites.
   d. Thickened chest wall

F. Pulmonary arterial wall calcification
1. Atherosclerotic calcification of the pulmonary artery wall occurs only in severe long-standing PAH.
2. Seen as thin arcuate calcification along the periphery of the enlarged central pulmonary arteries

G. Pulmonary artery thrombosis
1. Is generally felt to be an indication of chronic repeated pulmonary embolism as the cause of PAH
2. Can occur in severe, long-standing PAH of any cause

H. The radiographic findings in PAH from supravalvular/peripheral pulmonic stenosis are sometimes different as discussed above.

References and suggested additional reading

1. Miller SW. *Cardiac Radiology: The Requisites*. St. Louis: Mosby, 1996.
2. Müller NL, Fraser RS, Colman NC, Paré PD. *Radiologic Diagnosis of Diseases of the Chest*. Philadelphia: Saunders, 2001.
3. Simon M. Physiologic considerations in radiology of the pulmonary vasculature. In: Abrams HL ed. *Abrams Angiography Vascular and Interventional Radiology 3rd ed*. Boston: Little, Brown, 1983:783-802.
4. Steiner RM, Reddy GP, Flicker S. Congenital cardiovascular disease in the adult patient. J Thorac Imaging 2002; 17:1-17.
5. Woodring JH, Phillips BA, West JW, Ulmer J, Cooper JK. A prospective evaluation of plain radiographic signs of chronic obstructive pulmonary disease. J Thorac Imaging 1991; 6:14-21.

Chapter 7.

Decreased vascularity

I. **Radiographic signs of decreased pulmonary vascularity (oligemia)**
   A. **The hila are small.**
   B. **Visible pulmonary vessels in the lungs are small.**
      1. The decrease in vessel caliber may not be apparent if the oligemia is mild.
      2. The decrease in vessel caliber becomes more noticeable as the severity of oligemia increases.
   C. **Few if any vessels are visible through the liver.**
   D. **Lung density may be decreased due to the overall decrease in blood content of the lung.**
      1. The decrease in lung density may not be apparent on plain films if the oligemia is mild or the patient is small.
      2. The decrease in lung density becomes more noticeable as the severity of oligemia increases.
      3. CT is superior to plain films in demonstrating the decrease in pulmonary parenchymal density associated with oligemia.

II. **Decreased pulmonary vascularity — causes**
   A. **Congenital mechanical obstruction to pulmonary blood flow**
   B. **Pericardial tamponade**
   C. **Dehydration**
   D. **Addisonian crisis**
   E. **Pulmonary embolism**
   F. **Emphysema**

III. **Congenital mechanical obstruction to pulmonary flow**
   A. **With right-to-left (R-L) shunt at the atrial level**
      1. Severe pulmonic stenosis or pulmonic atresia with an intact ventricular septum (hypoplastic right heart syndrome, trilogy of Fallot)
         a. 0.3% of cases of congenital heart disease (CHD)
         b. There is critical stenosis or atresia of the pulmonic valve.
         c. The right ventricle (RV) is hypoplastic.
         d. There is a R-L shunt through an atrial septal defect (ASD) that allows blood to reach the left side of the heart.
         e. The right atrium (RA) is enlarged secondary to increased volume and pressure in the RA. This results in cardiomegaly.
         f. Blood flow to the lungs is primarily via a patent ductus arteriosus (PDA).
         g. Radiographic findings

        1) Decreased vascularity

        2) Enlarged cardiac silhouette secondary to right atrial enlargement (RAE)

        3) Because of the severe pulmonic stenosis or atresia the main pulmonary artery (MPA) segment is concave.

        4) Left aortic arch

    h. Treatment

        1) Balloon valvuloplasty for pulmonic stenosis

        2) Fontan procedure — placement of a conduit between the RV and pulmonary artery for those with pulmonic atresia

2. Tricuspid atresia

    a. 1.2% of cases of CHD

    b. Second most common cause of profound cyanosis in an infant after transposition of the great vessels (TGV)

    c. The tricuspid valve is absent and is replaced by a solid bar of fibrous tissue and fat.

    d. Classification of tricuspid atresia

        1) Type I — tricuspid atresia with normally related great vessels — 70-80% of cases

            a) With pulmonic atresia and an intact ventricular septum

                i) The RV is hypoplastic.

                ii) The RA is enlarged and there is a R-L shunt through an ASD.

                iii) Pulmonary blood flow is totally dependent upon a PDA.

                iv) Radiographic findings include decreased vascularity, cardiomegaly with RAE, and a concave MPA segment.

            b) With pulmonic stenosis and a small ventricular septal defect (VSD)

                i) The RV is hypoplastic.

                ii) The RA is enlarged and there is a R-L shunt through an ASD.

                iii) Blood reaches the lungs via either a PDA or VSD.

                iv) Radiographic findings include decreased vascularity, cardiomegaly with RAE, and a concave MPA segment. Serial diminution of pulmonary blood flow suggests that the VSD is spontaneously closing.

            c) With no pulmonic stenosis and a large VSD

                i) Presents with overcirculation of the pulmonary vascularity

                ii) Pulmonary flow and pressure are high and pulmonary vascular resistance can be increased (Eisenmenger's syndrome).

                iii) See Chapter 5.

        2) Type II — tricuspid atresia with D-TGV — 20% of cases

            a) With pulmonic atresia and a large VSD

> i) The aorta arises from the RV and the MPA arises from the left ventricle (LV). The MPA is hypoplastic because of the pulmonic atresia.
>
> ii) The RA is enlarged and there is a R-L shunt through an ASD to the left atrium (LA).
>
> iii) Blood flows from the LA to the LV and then through the VSD to the RV and out the aorta. Pulmonary blood flow is dependent upon a PDA.
>
> iv) Radiographic findings include decreased vascularity, cardiomegaly with RAE, and a concave MPA segment.

- b) With pulmonic stenosis and a large VSD
  - i) The aorta arises from the RV and the MPA arises from the LV.
  - ii) The RA is enlarged and there is a R-L shunt through an ASD to the LA.
  - iii) Blood flows from the LA to the LV and then through the VSD to the RV and out the aorta. Blood reaches the lungs via the PDA and from the LV through the stenotic pulmonic valve.
  - iv) Radiographic findings include decreased vascularity, cardiomegaly with RAE, and a concave MPA segment; however, if the pulmonic stenosis is mild, the pulmonary vascularity can be increased.
- c) With no pulmonic stenosis and a large VSD
  - i) The MPA arises from the LV. Since there is no pulmonic stenosis there is marked overcirculation of the pulmonary vascularity.
  - ii) Pulmonary flow and pressure are high and pulmonary vascular resistance can be increased (Eisenmenger's syndrome).
  - iii) See Chapter 5.
- 3) Type III — tricuspid atresia with L-TGV — rare

e. Overall, about 80% of cases of tricuspid atresia have decreased pulmonary vascularity and about 20% have overcirculation of the pulmonary vascularity.

f. Most cases have a left aortic arch, but mirror-image right aortic arch occurs in a few cases.

g. Tricuspid atresia is occasionally associated with other anomalies including:
  1) Musculoskeletal and gastrointestinal anomalies
  2) Down syndrome
  3) Asplenia
  4) Christmas disease (hemophilia B)
  5) Cat-eye syndrome

h. Treatment

1) A Blalock-Taussig shunt (side-to-side anastomosis of the subclavian artery and pulmonary artery on the side opposite the aortic arch) may be done initially for palliation.
    a) When a Blalock-Taussig shunt is done the ipsilateral arm is devascularized and the majority of blood flow to that extremity then comes from intercostal artery collaterals.
    b) When a Blalock-Taussig shunt is done on the right the patient may develop unilateral inferior rib notching on the right; conversely, when a Blalock-Taussig shunt is done on the left the patient may develop unilateral inferior rib notching on the left side.
2) A Glenn procedure (anastomosis of superior vena cava to right pulmonary artery) can also be done for palliation.
3) Definitive correction consists of a modified Fontan procedure (conduit between the RA and pulmonary artery) and closure of the ASD and VSD.
4) Increased pulmonary resistance from Eisenmenger's syndrome is a contraindication to the Fontan procedure

3. Ebstein's anomaly
   a. Uncommon anomaly characterized by downward displacement of the tricuspid valve into the RV. The anterior leaflet of the tricuspid valve is usually normally attached to the tricuspid annulus at the atrioventricular junction, but the posterior and septal leaflets are small and are downwardly displaced being attached to the ventricular wall below the annulus. This results in a large portion of the RV being "atrialized." The trabecular RV outflow chamber is small and ineffective at pumping blood, and serves as an obstruction to blood flow to the lungs.
   b. The papillary muscles and chordae tendineae are usually malformed.
   c. The abnormal formation of the tricuspid valve, papillary muscles, and chordae tendineae results in massive tricuspid regurgitation.
   d. The obstruction of blood flow through the RV and massive tricuspid regurgitation combine to produce massive dilatation of the RA. The superior vena cava (SVC), azygos vein, and inferior vena cava (IVC) are often dilated as well because of the massive tricuspid regurgitation.
   e. A R-L shunt occurs through an ASD.
   f. Ebstein's anomaly has been associated with maternal lithium use during pregnancy although the risk ratio remains unclear.
   g. Clinical manifestations
      1) Variable depending upon severity of malformation
      2) About 50% of cases present with cyanosis and right heart failure in infancy.
      3) The remaining cases present later in childhood with a murmur or abnormal chest radiograph but no symptoms, or present in late childhood or adult life with progressive dyspnea on exertion.

    h. Radiographic findings
       1) Decreased vascularity
       2) Cardiomegaly with RAE which can occasionally be massive
       3) Concave MPA segment. In some cases this results in the top of the left side of the cardiac silhouette being flattened. This is why the heart is said to be "box-shaped" in Ebstein's anomaly, although it takes quite a vivid imagination to see the heart shape as that of an actual "box."
       4) Tricuspid regurgitation may produce dilatation of the SVC, azygos vein, and IVC.
    i. Treatment
       1) A Blalock-Taussig shunt may be done initially for palliation. If hypoxemia remains a problem, a Glenn procedure may then be done.
       2) Modified Fontan procedure — closure of ASD and placement of a conduit between the RA and pulmonary artery
       3) Closure of the ASD and tricuspid valve reconstruction or replacement has become the most common form of treatment.
  4. Uhl's anomaly (parchment RV)
    a. Extremely rare
    b. Congenital hypoplasia of RV muscle — the epicardium and endocardium are present but the RV myocardium is completely absent
    c. Results in paper thin RV that is ineffective in pumping blood and becomes massively dilated.
    d. There is RAE with a R-L shunt through an ASD.
    e. Decreased vascularity
    f. Concave MPA segment

B. With R-L shunt at the ventricular level
  1. Tetralogy of Fallot (TOF)
    a. 8% of cases of CHD
    b. Most common cause of CHD associated with cyanosis and the most common form of cyanotic heart disease beyond infancy
    c. Consists of four components
       1) VSD — large and located immediately below the aortic valve
       2) Pulmonic stenosis — usually involves the infundibulum or right ventricular outflow tract (RVOT) and is caused by hypoplasia or hypertrophy of the crista supraventricularis
         a) Pulmonic valvular stenosis can also occur from either a bicuspid or unicuspid valve. This can contribute to the overall pulmonic stenosis but is rarely the only site of stenosis.
         b) The pulmonic valvular annulus and pulmonary arterial tree may be hypoplastic.
       3) Overriding aorta — the aorta straddles the VSD and receives blood from both ventricles
       4) Right ventricular hypertrophy (RVH)
    d. Associated anomalies

1) 25-30% have associated right aortic arch — almost always mirror-image branching
2) Anomalous origin of the left anterior descending coronary artery from the right coronary artery in 10%
   a) Unimportant physiologically
   b) The course of this vessel across the RVOT makes the usual site of right ventriculotomy and patch graft unavailable during surgery and requires use of a conduit to "jump over" the vessel.
3) ASD (pentalogy of Fallot) — 15%
4) PDA
5) Persistent left superior vena cava
6) Absence of the pulmonic valve with pulmonic regurgitation
7) Atresia of the pulmonic valve (pseudotruncus)
8) Atresia (interruption) or stenosis of the left pulmonary artery
9) Systemic arterial collateral blood supply
10) Ectopic origin of the left pulmonary artery from the ascending aorta (left hemitruncus)

e. TOF — radiographic findings
   1) The pulmonary vascularity is characteristically decreased. If the pulmonic stenosis is mild, the decrease in vascularity may be hard to appreciate on plain films. The combination of cyanosis and normal to slightly decreased vascularity indicates TOF.
   2) Concave MPA segment
   3) The cardiac size is typically normal. There is RVH that causes the cardiac apex to be elevated. The combination of the concave MPA segment and elevated cardiac apex results in the *coeur en sabot* appearance ("boot-shaped" heart) of TOF.
   4) Mirror-image right aortic arch in 25-30%
   5) If there is atresia of the left pulmonary artery, the left hilum and pulmonary vessels in the left lung may be quite small and there may be systemic arterial collateral flow to the left lung.
   6) If there is pulmonic atresia (pseudotruncus) there will be systemic arterial collateral flow to both lungs.

f. "Pink" TOF
   1) The abnormalities are the same as regular TOF except that the pulmonic stenosis is very mild.
   2) There is overcirculation of the pulmonary vascularity with a left-to-right shunt through the VSD.
   3) Clinically and radiographically resembles simple VSD

g. Treatment
   1) Palliative treatment, when done, now consists of a modified Blalock-Taussig shunt with placement of a graft between the subclavian artery and ipsilateral pulmonary artery on the side opposite the aortic arch.
   2) Corrective surgery for patients with pulmonic stenosis consists of closing the VSD through a right ventriculotomy, resecting

infundibular muscle, and, if the infundibulum, pulmonic valve, and MPA are hypoplastic, using a pericardial patch graft to open the narrowed area. If the patient has pentalogy of Fallot, the ASD is also closed.

      3) Corrective surgery for those with pulmonic atresia who have good-sized pulmonary arteries consists of closing the VSD and placing a conduit between the right ventricle and pulmonary artery (Fontan procedure)

  2. Other conditions which may have both a VSD and severe pulmonic stenosis or atresia and therefore may mimic TOF
    a. TGV
    b. Corrected TGV
    c. Single ventricle
    d. Double outlet right ventricle

C. Summary — radiographic signs of cyanotic CHD associated with decreased vascularity
  1. R-L shunt at the atrial level
    a. Decreased vascularity
    b. Cardiomegaly with RAE
    c. Concave MPA segment
    d. Left aortic arch most common
  2. R-L shunt at ventricular level
    a. Decreased vascularity
    b. Normal sized heart with normal RA
    c. Concave MPA segment
    d. R aortic arch common
  3. To simplify this even further, if the pulmonary vascularity is decreased and the heart is enlarged the differential diagnosis includes severe pulmonic stenosis or atresia with an intact ventricular septum, tricuspid atresia, Ebstein's anomaly, and Uhl's anomaly. If the pulmonary vascularity is decreased or near normal and the heart size is normal the patient most likely has TOF.

IV. Pericardial tamponade
  A. Oligemia
    1. The pulmonary vascularity is often normal.
    2. Decreased atrial filling, however, can lead to oligemia that can be severe.
  B. Enlarged cardiac silhouette
    1. Due to pericardial effusion
  C. Widened vascular pedicle
    1. Due to systemic venous hypertension

V. Dehydration or Addisonian crisis
  A. Oligemia
    1. Due to decreased circulating blood volume

B. Small heart
1. Due to decreased circulating blood volume
C. Narrow vascular pedicle
1. Due to decreased circulating blood volume

VI. Pulmonary embolism (PE)
A. Acute PE
1. The chest radiograph may be completely normal in 10-15% of cases. Air-space consolidation, linear atelectasis, and pleural effusion may be seen but are signs with both low-sensitivity and low-specificity. The major role of the plain chest radiograph is in excluding other causes for the patient's symptoms such as pneumothorax or lobar or whole lung collapse.
2. Radiographic signs with low-sensitivity but high-specificity for PE.
   a. Oligemia
      1) May be focal — Westermark's sign
      2) Can be diffuse and bilateral in cases of massive pulmonary embolism
   b. Enlargement of the descending branch of the right or left pulmonary artery
      1) Fleischner's sign — the descending branch of the right or left pulmonary artery is enlarged but tapers normally
      2) The "knuckle" sign — the descending branch of the right or left pulmonary artery is enlarged and terminates abruptly at the lobar arterial level
   c. Hampton's hump
      1) Peripheral, pleural-based, wedge-shaped or convex opacity abutting either the chest wall, diaphragm, or an interlobar fissure with its apex directed toward the hilum.
      2) Represents a pulmonary infarct.
      3) Has high-specificity for PE in the correct clinical setting but is uncommon and therefore has low-sensitivity.
   d. Signs of acute pulmonary arterial hypertension (PAH)
      1) Enlarged MPA, enlarged central pulmonary arteries, and enlarged right heart
      2) Signs of acute PAH are uncommon in acute PE but can be seen in cases of massive bilateral PE and large obstructing central saddle emboli.
      3) There may be associated signs of right heart failure.
3. Spiral, helical, or electron-beam CT
   a. Vascular findings
      1) Intravascular filling defect forming acute angles with the vessel wall and surrounded by contrast
      2) Complete cut-off of vascular opacification
      3) Increased diameter of occluded vessel
   b. Parenchymal findings

          1) Decreased parenchymal attenuation and vascularity (oligemia) in areas of lung supplied by pulmonary arteries containing filling defects

          2) Non-enhancing peripheral, pleural-based, wedge-shaped or convex opacities consistent with pulmonary infarcts

          3) Linear atelectasis

      c. Sensitivity 80-90%, specificity 80-95%

          1) False-negative interpretations mainly due to subsegmental PE

          2) False-positive interpretations mainly due to hilar nodes and technical problems

B. Chronic PE

    1. Chest radiographic findings are typically those of PAH

    2. Spiral, helical, or electron-beam CT

      a. Vascular findings

          1) Eccentric filling defects contiguous with the vessel wall (thrombus)

          2) Recanalization of filling defect (concentric thrombus with central recanalized lumen opacified by contrast)

          3) Arterial stenosis or webs

          4) Reduction of more than 50% in arterial diameter

          5) Abrupt cut-off of arterial branches

      b. Parenchymal findings

          1) Alternating localized areas of decreased parenchymal attenuation and vascularity that are sharply marginated from adjacent areas of increased attenuation and vascularity (mosaic oligemia or mosaic perfusion)

          2) Non-enhancing peripheral, pleural-based, wedge-shaped or convex opacities consistent with pulmonary infarcts

      c. Signs of PAH

    3. Although signs of pulmonary arterial thrombosis in patients with PAH are generally considered to indicate chronic, recurrent PE as the cause of PAH, pulmonary arterial thrombosis can occur in severe, long-standing PAH of any cause.

VII. Emphysema

  A. Emphysema is the most common cause of pulmonary oligemia in an adult.

  B. Distribution of emphysema

    1. Centrilobular emphysema is characteristically upper lobe predominant in distribution.

    2. Panlobular emphysema is characteristically basilar predominant in distribution.

    3. Emphysema, however, can be diffuse and involve all zones of both lungs equally. In addition, the distribution of emphysema can be patchy and random with asymmetric areas of involvement in both lungs.

  C. Radiographic signs of emphysema

    1. Signs of hyperinflation

      a. Low-positioned diaphragm — present if the highest portion of the right hemidiaphragm projects below the inferior cortical margin of the right 11th rib

      b. Increased anteroposterior diameter of the thorax

      c. Deepened retrosternal clear space

      d. Widened costophrenic angle — determined by deepening and widening of one or both of the lateral costophrenic angles with a rounded, rather than sharp, angular inferior margin

      e. Flattened or inverted hemidiaphragms

  2. Signs of parenchymal lung destruction

      a. Oligemia — determined by decreased diameter, stretching, and diminished branching of pulmonary vessels with resultant hyperlucency

      b. Organic redistribution — increased vascularity in areas of lung less severely involved by emphysema

        1) In upper lobe predominant emphysema the pulmonary vascularity in the lung bases is increased.

        2) In basilar predominant emphysema the pulmonary vascularity in the upper lobes will be increased.

        3) In cases of randomly distributed emphysema there will be increased pulmonary vascularity in the more normal areas of lung between the more emphysematous areas.

      c. Bulla formation

  3. Signs of associated chronic bronchitis

      a. Thickened bronchial walls seen end-on

      b. Tram-line shadows

      c. Accentuated, irregular bronchovascular markings

  4. Signs of associated PAH

      a. Enlargement of the MPA

      b. Enlargement of the descending branches of the right and left pulmonary arteries to 2 cm or greater

      c. Enlargement of the right-sided cardiac chambers

D. Differentiation of emphysema from cardiac disease

  1. In oligemia due to cardiac disease the pulmonary vessels are small but they branch normally. Oligemia due to emphysema can be distinguished from that caused by cardiac disease by the fact that in emphysema the vessels are thinned, stretched, and demonstrate diminished branching. Organic redistribution and bulla formation do not occur in cardiac disease, and indicate that the decreased vascularity is secondary to emphysema. Associated signs of hyperinflation and chronic bronchitis also aid in this distinction.

  2. Emphysema is a common and important cause of PAH. Again, signs of hyperinflation, parenchymal lung destruction, and associated chronic bronchitis should aid in the determination that PAH is secondary to emphysema rather than other conditions.

3. The diagnosis of left ventricular failure in patients with emphysema can be difficult. This subject is reviewed in detail in Chapter 4.
4. Emphysema alters pulmonary ventilation and perfusion and characteristically causes matched defects on radionuclide ventilation/perfusion imaging. These defects hinder the detection of PE on ventilation/perfusion scintigraphy and may render the examination indeterminate for PE. CT may be required in these patients to determine whether or not they have PE.

References and suggested additional reading

1. Dähnert W. *Radiology Review Manual 4th ed.* Philadelphia: Lippincott Williams & Wilkins, 2000.
2. Elliott LP. Pathophysiology and roentgenologic findings in Ebstein's malformation of the tricuspid valve. In: Taveras JM, Ferrucci JT eds. *Radiology: Diagnosis-Imaging-Intervention.* Volume 2, Chapter 114. Hagerstown: Lippincott, 1991.
3. Fuster V, Alexander RW, O'Rourke RA, Roberts R, King SB III, Wellens HJJ. *Hurst's The Heart 10th ed.* New York: McGraw-Hill, 2001.
4. Gedgaudas E, Moller JH, Castaneda-Zuniga WR, Amplatz K. *Cardiovascular Radiology.* Philadelphia: Saunders, 1985.
5. Meszaros WT. *Cardiac Roentgenology: Plain Films and Angiocardiographic Findings.* Springfield: Charles C Thomas, 1969.
6. Miller SW. *Cardiac Radiology: The Requisites.* St. Louis: Mosby, 1996.
7. Müller NL, Fraser RS, Colman NC, Paré PD. *Radiologic Diagnosis of Diseases of the Chest.* Philadelphia: Saunders, 2001.
8. Spindola-Franco H, Fish BG. *Radiology of the Heart: Cardiac Imaging in Infants, Children, and Adults.* New York: Springer-Verlag, 1985.
9. Woodring JH, Phillips BA, West JW, Ulmer J, Cooper JK. A prospective evaluation of plain radiographic signs of chronic obstructive pulmonary disease. J Thorac Imaging 1991; 6:14-21.

Chapter 8.

Systemic arterial blood supply to the lungs

I. **The causes of systemic arterial blood supply to the lungs can be broken down into two main categories.**
    A. **Systemic arterial blood supply to the lungs that represents collateral blood flow to the lungs in patients with congenital or acquired obstruction to pulmonary blood flow**
        1. Congenital cardiac disorders associated with pulmonic atresia
        2. Congenital absence or interruption of the proximal portion of a pulmonary artery
        3. Mediastinal fibrosis with occlusion of a pulmonary artery
    B. **Anomalous systemic arterial blood supply to the lungs that is part of a congenital developmental anomaly of the lungs themselves and that is not associated with either congenital or acquired obstruction to pulmonary blood flow.**
        1. Pulmonary sequestration
        2. Systemic arterialization of the lung without sequestration

II. **Systemic arterial blood supply to the lungs that represents collateral blood flow to the lungs in patients with congenital or acquired obstruction to pulmonary blood flow**
    A. **Congenital cardiac disorders associated with pulmonic atresia**
        1. Tetralogy of Fallot (TOF) with atresia of the pulmonic valve (pseudotruncus)
            a. Pseudotruncus accounts for 99% of all cases of congenital heart disease associated with systemic arterial collateral flow to the lungs.
            b. The pulmonic valve is atretic and the central right and left pulmonary arteries and hilar arteries are small.
            c. Large systemic arterial collaterals arise from the descending thoracic aorta and connect to the right and left pulmonary arteries to reconstitute blood flow to the lungs. These large arteries represent greatly enlarged bronchial arteries. Transpleural collaterals from the intercostal arteries can also develop but the major collateral pathway for pulmonary blood flow in pseudotruncus is through the bronchial arteries.
            d. The other features of TOF are still present
                1) Ventricular septal defect (VSD)
                2) Overriding aorta that receives blood from both ventricles
                3) Right ventricular hypertrophy (RVH)

e. The condition is called pseudotruncus because only one great artery, the ascending aorta, arises from the heart and therefore the condition superficially resembles truncus arteriosus. Pseudotruncus can be distinguished from true truncus arteriosus by determining the origin of the pulmonary arteries.

   1) In truncus arteriosus the pulmonary arteries arise from the truncus.

   2) In pseudotruncus the pulmonary arteries arise via large systemic collaterals from the descending thoracic aorta.

f. Pseudotruncus — radiographic findings

   1) Concave main pulmonary artery (MPA) segment

   2) The cardiac size is typically normal. There is RVH that causes the cardiac apex to be elevated. The combination of the concave MPA segment and elevated cardiac apex results in the *coeur en sabot* appearance ("boot-shaped" heart) similar to TOF.

   3) Mirror-image right aortic arch is also common as in TOF.

   4) The hila are typically small but may be normal-sized or enlarged.

      a) Even though pulmonary blood flow is reconstituted via systemic arterial collaterals that connect to the pulmonary arteries, the overall pulmonary blood flow is still usually diminished and the hila usually are small.

      b) In some cases of pseudotruncus the bronchial arterial collaterals are so large that the hila may be normal-sized or even enlarged.

   5) The visible pulmonary vessels in the lungs are "bizarre" in appearance.

      a) There are two major pathways of collateral blood flow from the systemic arteries to the pulmonary arteries in pseudotruncus.

         i) Bronchial artery to proximal pulmonary artery collaterals

         ii) Intercostal artery to peripheral pulmonary artery collaterals

      b) Bronchial artery to proximal pulmonary artery collaterals

         i) There are microscopic communications between the bronchial arteries and pulmonary arteries that are present but not functioning in normal subjects.

         ii) In disease states that result in occlusion of pulmonary blood flow, these anastomoses open up allowing the bronchial arteries to supply blood to the pulmonary arterial circulation.

         iii) Bronchial artery collaterals arising from the descending thoracic aorta typically communicate with the proximal pulmonary arteries.

         iv) Bronchial arterial collaterals are often nodular and serpiginous or even "cork-screw" in appearance. They do not follow a normal course through the lung and it may be possible to follow them to the descending aorta on plain films. CT and MR are excellent means of identifying these collateral vessels.

      c) Intercostal artery to peripheral pulmonary artery collaterals

i) Normally no communications exist between the intercostal arteries and the peripheral pulmonary arteries.

ii) In conditions that produce severe obstruction to pulmonary blood flow for a prolonged period of time, usually for at least several years, systemic collateral arteries will arise from the intercostal arteries, pierce the parietal pleura of the chest wall, cross the pleural space, pierce the visceral pleural covering of the lung, and communicate with the peripheral pulmonary arteries in order to help supply blood to the lung. The exact mechanism by which transpleural collateral arteries develop is unknown but is probably related to stimulation of stem cells by chronic severe hypoxia. In pseudotruncus, transpleural collaterals are generally not present in infancy but they will develop over time if the patient is untreated.

iii) Transpleural collaterals occur in untreated pseudotruncus, congenital absence or interruption of the proximal portion of a pulmonary artery, and mediastinal fibrosis with occlusion of a pulmonary artery.

iv) Transpleural collaterals coming from the intercostal arteries are often mistaken for interstitial lung disease. They are most prominent in the periphery of the lung and are usually more pronounced in the lung apex than at the base. They have a serpiginous almost "reticulonodular" appearance and can even be somewhat linear in configuration resembling Kerley A or B lines from thickening of the interlobular septa.

v) Although these transpleural collateral arteries are well demonstrated by plain films, CT, and MR, it is difficult to demonstrate the communications between these transpleural collateral arteries and the peripheral pulmonary arteries by either plain films or MR. CT, however, usually nicely demonstrates the serpiginous transpleural collateral arteries connecting to the peripheral pulmonary arteries.

6) The radiographic density of the lungs is often altered.

a) Typically the lungs demonstrate decreased density due to oligemia.

b) In lungs chronically subjected to systemic arterial collateral blood flow pulmonary density may be increased because of three factors.

i) The presence of the collateral vessels themselves

ii) Pulmonary edema due to the fact that the pulmonary arteries are subjected to systemic hydrostatic pressure by way of the collateral arteries

iii) Repeated episodes of pulmonary hemorrhage from the systemic collateral arteries resulting in pulmonary hemosiderosis and fibrosis

          7) If transpleural collateral arteries are well-developed, there may be bilateral inferior rib notching.

    g. Pseudotruncus — treatment

          1) Corrective surgery for those with pulmonic atresia who have good-sized pulmonary arteries consists of closing the VSD and placing a conduit between the right ventricle and pulmonary artery (Fontan procedure).

          2) For those patients whose pulmonary arteries are severely hypoplastic or discontinuous, balloon dilatation of hypoplastic areas and anastomosis of discontinuous vessels is performed initially in the hope that corrective surgery with a conduit and closure of the VSD may then be possible.

2. TOF with atresia of the left pulmonary artery

    a. If there is atresia of the left pulmonary artery, the left hilum and pulmonary vessels in the left lung may be quite small and there may be systemic arterial collateral flow to the left lung.

    b. There may be associated unilateral inferior rib notching on the left side.

3. Others

    a. Tricuspid atresia, transposition of the great vessels, double-outlet right ventricle, and single ventricle can be associated with pulmonic atresia and can have systemic arterial blood flow to the lungs.

    b. All of these conditions combined account for less than 1% of cases of congenital heart disease associated with systemic arterial collateral blood supply to the lungs.

B. Congenital absence or interruption of the proximal portion of a pulmonary artery

1. Is due to interruption of the proximal portion of the primitive right or left sixth aortic arch

2. Congenital absence or interruption (atresia) of the left pulmonary artery usually occurs in association with TOF (see above).

3. Congenital absence or interruption of the right pulmonary artery (RPA) usually occurs as an isolated abnormality although there is a slight increase in the incidence of acyanotic left-to-right shunts and congenital abnormalities of the aortic arch.

    a. In the older literature this entity was referred to as congenital "absence" of the RPA.

    b. The interlobar or descending branch of the RPA, lobar branches of the RPA and branches of the RPA distal to the lobar arteries are present. However, since the central portion of the RPA that would normally connect these branches to the MPA is absent, the more recent literature refers to this entity as congenital "interruption" of the proximal RPA.

    c. The bronchial arteries and intercostal arteries serve as the main sources of collateral blood supply to the right lung.

    d. Radiographic findings

1) The right lung is usually hypoplastic and smaller than the left.
2) The right hilum is small.
3) The right lung is often of decreased density due to oligemia.
4) The left hilum and pulmonary vessels in the left lung are enlarged since the left lung receives the entire right ventricular output.
5) In infancy and early childhood collateral blood flow comes primarily from the bronchial arteries. Bronchial arterial collaterals may be seen as serpiginous or "cork-screw" vessels near the right hilum, but they are often relatively unapparent and the radiographic findings are often limited to a small, oligemic right lung with a small hilum and a left lung that has a prominent hilum and increased vascularity. Transpleural collaterals from the intercostal arteries begin to appear radiographically during late childhood and increase in prominence as time goes by. These transpleural collaterals are most prominent in the periphery of the lung and are usually more pronounced in the lung apex than at the base. They have a serpiginous almost "reticulonodular" appearance and can even be somewhat linear in configuration resembling Kerley A or B lines. Lung density may be increased because of the collateral vessels themselves, pulmonary edema due to the fact that the pulmonary arteries are subjected to systemic hydrostatic pressure by way of the collateral arteries, and repeated episodes of pulmonary hemorrhage from the systemic collateral arteries resulting in pulmonary hemosiderosis and fibrosis.
6) Unilateral inferior rib notching on the right side may also occur.
7) While CT and MR are equally good at demonstrating the missing central portion of the RPA, CT is better at demonstrating the collateral vessels, particularly the transpleural collaterals, and their anastomoses with the pulmonary arteries.

4. Most patients are asymptomatic. In some cases pulmonary arterial hypertension and congestive heart failure develop due to the chronic overcirculation of the left lung. Patients with interruption of the RPA may also experience hemoptysis due to bleeding from the systemic arterial collateral arteries.

C. **Mediastinal fibrosis with occlusion of a pulmonary artery**
   1. Rare condition characterized by chronic inflammation and fibrosis of mediastinal soft tissues.
      a. Can be diffuse
      b. Can be focal and "mass-like"
      c. Often associated with compression and occasionally occlusion of one or more of the pulmonary arteries or veins, the superior vena cava, the central bronchi, or esophagus
   2. Mediastinal fibrosis is usually secondary to histoplasmosis. Other causes include tuberculosis, methysergide therapy, mediastinal irradiation, autoimmune disease, and idiopathic.
   3. Characteristic radiologic findings include

    a. The diffuse form is associated with extensive widening of the mediastinum particularly of the upper mediastinum. Calcification is uncommon even when histoplasmosis is the cause.

    b. The focal form is characterized by the presence of a mass.

       1) Usually involves the right paratracheal region or right hilum although the subcarinal area, left hilum, and posterior mediastinum can be involved

       2) Calcification is present in 60-90%.

    c. There is compression or occlusion of one or more of the pulmonary arteries or veins, the superior vena cava, the central bronchi, or esophagus.

4. When mediastinal fibrosis results in occlusion of a pulmonary artery, typically the RPA, systemic arterial collaterals develop from both the bronchial arteries and intercostal arteries.

    a. The involved lung, typically the right lung, is smaller than the contralateral lung.

    b. The hilum is enlarged and there is usually a discrete hilar mass.

       1) The hilar mass is usually partially calcified.

       2) The hilar mass encases and occludes the ipsilateral pulmonary artery.

       3) The presence of a hilar mass distinguishes mediastinal fibrosis from congenital interruption of the pulmonary artery. Otherwise the two conditions are virtually identical.

    c. The contralateral hilum and pulmonary vessels, typically on the left side, are enlarged since the contralateral lung receives the entire right ventricular output.

    d. Bronchial arterial collaterals may be seen as serpiginous or "corkscrew" vessels near the right hilum, and transpleural intercostal arterial collaterals in the periphery of the lung have a serpiginous almost "reticulonodular" appearance and can even be somewhat linear in configuration resembling Kerley A or B lines.

    e. Lung density is often increased because of the collateral vessels themselves, pulmonary edema due to the fact that the pulmonary arteries are subjected to systemic hydrostatic pressure by way of the collateral arteries, and repeated episodes of pulmonary hemorrhage from the systemic collateral arteries resulting in pulmonary hemosiderosis and fibrosis.

    f. Unilateral inferior rib notching on the ipsilateral side may also occur.

    g. While CT and MR are equally good at demonstrating the hilar mass obstructing the RPA, CT is better at demonstrating calcification within the mass and is, therefore, the preferred imaging modality for evaluating mediastinal fibrosis.

5. Clinical signs and symptoms depend on which mediastinal structures are involved. Patients may be asymptomatic, or may present with signs and symptoms of superior vena cava syndrome, wheezing, and dysphagia. Recurrent hemoptysis, due to bleeding from systemic arterial collateral

vessels, is the most common symptom related to obstruction of a pulmonary artery.

6. One might question how focal mediastinal fibrosis in the hilar area with occlusion of a pulmonary artery can be distinguished radiographically from a central hilar carcinoma occluding a pulmonary artery. The answer is the transpleural intercostal arterial collaterals. As previously mentioned, it takes years for these collateral vessels to develop. Undiagnosed, untreated patients with bronchogenic carcinoma occluding a central pulmonary artery do not live long enough for transpleural collaterals to develop. So, if you can identify transpleural collateral arteries from the intercostal arteries communicating with the peripheral pulmonary arteries, you can be fairly confident that the patient has mediastinal fibrosis rather than bronchogenic carcinoma.

III. **Anomalous systemic arterial blood supply to the lungs that is part of a congenital developmental anomaly of the lungs themselves and that is not associated with either congenital or acquired obstruction to pulmonary blood flow.**
  A. **Pulmonary sequestration**
    1. Two forms
        a. Intralobar sequestration
        b. Extralobar sequestration
    2. Intralobar sequestration
        a. Accounts for 75% of cases
        b. The sequestered lung tissue is contained within the visceral pleura of the normally functioning lung.
        c. Most cases involve the posterior basal segment of a lower lobe.
            1) Twice as common on the left as the right
            2) May be bilateral
        d. May rarely involve an upper lobe
        e. Arterial blood supply
            1) Arterial blood supply comes from anomalous systemic arteries arising from the descending thoracic aorta (most common), upper abdominal aorta, or celiac or splenic arteries.
            2) Anomalous arteries enter the lung through the inferior pulmonary ligament.
        f. Venous drainage
            1) In 95% of cases venous drainage is to the left atrium via the pulmonary veins.
                a) Creates a left-to-left (L-L) shunt
                b) This situation (L-L shunt) is unique to sequestration.
            2) Systemic venous drainage to the azygos, hemiazygos, or intercostal veins, or the inferior or superior vena cava occurs in the remaining cases.
            3) Rarely, the venous drainage from intralobar sequestration may be to the ipsilateral pulmonary artery.

g. Pathologic and radiographic findings
   1) Intralobar sequestration usually consists of immature but otherwise relatively normal pulmonary parenchyma.
   2) The rudimentary bronchial system within intralobar sequestration does not communicate with the tracheobronchial tree. Intralobar sequestration is usually aerated, nevertheless, presumably via collateral air drift from surrounding normal lung by way of the pores of Kohn. Intralobar sequestration may appear normally aerated on plain films; however, lung ventilated by collateral air drift usually develops postobstructive hyperinflation. This is best appreciated on CT but can occasionally be seen on plain films.
   3) Because the bronchi within the area of intralobar sequestration are obstructed, mucoid impaction of the bronchi may occur.
   4) Occasionally the anomalous systemic arteries supplying the intralobar sequestration can be seen on plain films as tubular or serpiginous structures coming from the descending aorta or region of the diaphragm and entering the sequestered lung. CT and MR are both excellent means of demonstrating the anomalous systemic arteries and have essentially replaced arteriography for this purpose. CT, however, is better than MR in showing all the features of the disease, particularly the postobstructive hyperinflation of the sequestered lung.
   5) The combination of mucoid impaction of the bronchus surrounded by hyperinflated lung is one characteristic feature of intralobar sequestration.
      a) The identification of systemic arterial blood supply to the hyperinflated lung would confirm the diagnosis.
      b) Bronchial atresia, which also typically has mucoid impaction of the bronchus surrounded by hyperinflated lung, can be distinguished from intralobar sequestration by the lack of systemic arterial blood supply to the hyperinflated lung.
   6) When intralobar sequestration becomes infected, it may appear as either a solid mass or an infiltrate containing air bronchograms. A history of repeated episodes of pneumonia in the same location should raise the suspicion of possible intralobar sequestration.
   7) Rare manifestations of intralobar sequestration
      a) Intralobar sequestration may present as a calcified mass.
      b) Intralobar sequestration may present as a cystic mass consisting of one or more cysts of varying size, which may contain air-fluid levels.

3. Extralobar sequestration
   a. Accounts for 25% of cases
   b. The sequestered lung tissue is typically contained within a distinct visceral pleural coat and usually maintains complete anatomical and functional separation from the normal lung.

    c. Extralobar sequestration is most often found between the lower lobe and diaphragm but may be found in a number of locations, including within the substance of the diaphragm, the lung, the hilum, the mediastinum, the pleural or pericardial spaces, the retroperitoneum, or the upper peritoneal space.

    d. Extralobar sequestration is much more common on the left side than the right.

    e. Arterial blood supply
        1) In most cases the arterial blood supply comes from anomalous systemic arteries arising from the aorta.
        2) Anomalous arteries may also come from the splenic, gastric, subclavian, and intercostal arteries.
        3) Rarely, the blood supply may come from the pulmonary arteries.

    f. Venous drainage
        1) The venous drainage is usually systemic, with drainage occurring to the azygos vein, hemiazygos vein, inferior vena cava, or portal vein.
        2) Rarely, venous drainage may be to the pulmonary veins.

    g. Pathologic and radiographic findings
        1) Since extralobar sequestration seldom has a communication with the tracheobronchial tree or surrounding lung, it is usually seen as a solid mass, typically in the left paraspinal region or along the left hemidiaphragm.
        2) If a communication to the tracheobronchial tree or esophagus exists, extralobar sequestration may become infected and appear as a mixed cystic and solid mass.

    h. The diagnosis usually rests on the identification of the anomalous systemic arteries feeding the mass.

B. Systemic arterialization of the lung without sequestration
    1. This is the most rare of all causes of systemic arterial blood supply to the lungs.
    2. Systemic arterialization of the lung without sequestration occurs on the right side in association with congenital pulmonary venolobar syndrome (CPVS).
        a. Typically there is hypogenetic lung on the right.
            1) The right lung is small.
            2) The right hilum is small and the pulmonary vascularity of the right lung is diminished.
        b. There may be partially anomalous venous return (scimitar syndrome), duplication of the diaphragm (accessory diaphragm), and other features of CPVS including horseshoe lung.
        c. Part or all of the hypogenetic right lung may receive its arterial blood supply from the thoracic or abdominal aorta or their branches.
            1) In some cases of hypogenetic lung the anomalous systemic arteries supply an area of pulmonary sequestration.

       2) In most cases, the anomalous systemic arteries supply only the hypogenetic lung tissue.

3. Systemic arterialization of the lung without sequestration occurs on the left side as an isolated anomaly.
   a. Typically involves the left lower lobe (LLL)
   b. The left hilum is small because the normal LLL pulmonary artery is missing.
   c. A large serpiginous vessel arises from the descending thoracic aorta and enters the LLL to connect to normal appearing segmental pulmonary arteries in the basal segments of the LLL.
   d. Although the LLL is considered to be "normal" because there is no sequestration, the LLL is usually small and often demonstrates ground-glass opacification due to pulmonary edema or alveolar hemorrhage induced by the high systemic arterial pressure.
4. Several cases of systemic arterial blood supply to the right lower lobe (RLL) without sequestration have been reported in which the anomalous artery to the RLL arose from an anomalous systemic artery supplying an area of sequestration in the LLL.

References and suggested additional reading

1. Felson B. Pulmonary sequestration revisited. Med Radiogr Photogr 1988; 64:1-27.
2. Fuster V, Alexander RW, O'Rourke RA, Roberts R, King SB III, Wellens HJJ. *Hurst's The Heart 10th ed.* New York: McGraw-Hill, 2001.
3. Gedgaudas E, Moller JH, Castaneda-Zuniga WR, Amplatz K. *Cardiovascular Radiology*. Philadelphia: Saunders, 1985.
4. Kim TS, Lee KS, Im J-G, Jin MG, Park JS, Kim JH. Systemic arterial supply to the normal basal segments of the left lower lobe: radiographic and CT findings in 11 patients. J Thorac Imaging 2002: 17:34-39.
5. Meszaros WT. *Cardiac Roentgenology: Plain Films and Angiocardiographic Findings*. Springfield: Charles C Thomas, 1969.
6. Müller NL, Fraser RS, Colman NC, Paré PD. *Radiologic Diagnosis of Diseases of the Chest*. Philadelphia: Saunders, 2001.
7. Spindola-Franco H, Fish BG. *Radiology of the Heart: Cardiac Imaging in Infants, Children, and Adults*. New York: Springer-Verlag, 1985.
8. Woodring JH, Howard TA, Kanga JF. Congenital pulmonary venolobar syndrome revisited. RadioGraphics 1994; 14:349-369.

Chapter 9.

Abnormalities of cardiac position and visceral situs

I. **General rules in evaluation of cardiac position and visceral situs**
   A. **The position of the cardiac apex determines whether there is dextrocardia or levocardia.**
      1. If the cardiac apex is on the right, there is dextrocardia.
      2. If the cardiac apex is on the left, there is levocardia.
   B. **Situs refers to the position of the viscera and atria in the body.**
      1. Visceral situs corresponds to atrial situs.
         a. Situs solitus (normal situs) indicates that the right atrium (RA) is on the right, and the left atrium (LA) is on the left. The liver is on the right, and the stomach and spleen are on the left.
         b. Situs inversus is the mirror-image of normal thoracic and visceral situs. The anatomic RA is on the left, and the anatomic LA is on the right. The liver is on the left, and the stomach and spleen are on the right.
      2. The situs of the lungs (thoracic situs) corresponds to atrial situs.
   C. **The ventricles may or may not be concordant with their respective atria.**
   D. **The atrioventricular (AV) valves (tricuspid and mitral valves) stay with their normally related ventricle even when the ventricle is in the wrong location.**
      1. The tricuspid valve stays with the right ventricle (RV).
      2. The mitral valve stays with the left ventricle (LV).

II. **Abnormalities of cardiac position associated with normal visceral situs (situs solitus) and situs inversus**
   A. **Dextroposition and dextrocardia**
      1. Dextroposition with situs solitus
         a. The heart is abnormally positioned to the right because of an extrinsic factor, usually hypoplasia or agenesis of the right lung.
         b. The LV is still to the left of and posterior to the RV as it is in a normal heart and the cardiac apex is still directed toward the left.
         c. Visceral situs is normal (situs solitus).
         d. Dextroposition with situs solitus may be discovered incidentally in asymptomatic patients; however, there is an increased incidence of associated congenital anomalies.
            1) Left-to-right (L-R) intracardiac shunts
            2) Congenital pulmonary venolobar syndrome (CPVS)
               a) The major components of CPVS include hypogenetic lung, partially anomalous pulmonary venous return (PAPVR), congenital interruption of the right pulmonary artery, pulmonary

sequestration, systemic arterialization of the lung without sequestration, absence of the inferior vena cava (IVC), and accessory right hemidiaphragm. These abnormalities can occur alone or in almost any combination.

    b) Although CPVS can involve either lung, it is much more common on the right, with approximately 75% of cases being right-sided. The right lung is small and the heart is typically abnormally positioned to the right.

    c) The combination of hypogenetic lung and PAPVR is called scimitar syndrome. Scimitar syndrome is an almost exclusively right-sided condition. Part or all of the hypogenetic lung is drained by an anomalous pulmonary vein that does not drain to the LA. Instead, it drains to the systemic venous system (IVC below the diaphragm, suprahepatic portion of the IVC, hepatic veins, portal vein, azygos vein, coronary sinus) or occasionally the RA resulting in a L-R shunt. As the anomalous draining vein courses toward the diaphragm it casts a curvilinear shadow in the right lower lung zone that has been likened to a Turkish sword or scimitar. Occasionally several anomalous pulmonary veins may be present.

2. Dextrocardia with situs solitus (isolated dextrocardia)
   a. The heart is rotated such that its apex lies on the right side. The RV forms part of the right heart border. The LV also lies more to the right than normal, and its apex is also directed to the right. The ventricles lie side-by-side, the LV being situated to the left of the RV rather than behind it as usual. The main pulmonary artery (MPA) is medially situated and no longer forms part of the left heart border.
   b. The aortic knob, descending thoracic aorta, and stomach are on the left.
   c. There is a high incidence of associated cardiac anomalies (80-95%)
      1) Corrected transposition of the great vessels (L-TGV) — most common associated cardiac defect
      2) Complete transposition of the great vessels (D-TGV)
      3) Ventricular septal defect (VSD)
      4) Pulmonic stenosis
   d. Cyanosis is frequently present
3. Mirror-image dextrocardia (dextrocardia with situs inversus)
   a. There is a complete reversal of cardiac, pulmonary, and visceral anatomy resulting in an exact mirror-image of normal.
   b. The IVC, anatomic RA, and major lobe of the liver are on the left side. The cardiac apex, descending thoracic aorta, anatomic LA, and stomach are on the right.
   c. The right-sided lung has two lobes and its artery emanates from the mediastinum above the right-sided main bronchus resulting in a right-sided hyparterial main bronchus. The left-sided lung has three lobes,

and its artery emanates from the mediastinum below the left-sided main bronchus resulting in an eparterial main bronchus.

    d. Most patients are normal with no associated abnormalities; however, there is an increased incidence of associated congenital anomalies compared to true normals. About 5-10% of patients have associated congenital defects.

      1) Kartagener's syndrome

        a) Classic triad of dextrocardia with situs inversus, bronchiectasis, and chronic sinusitis

        b) The basic defect is absence of the dynein arms, small structures containing adenosine triphosphatase that link the microtubules of normal cilia together and that are necessary for normal ordered ciliary motility. In Kartagener's syndrome, the absence of the dynein arms results in markedly dyskinetic and disordered ciliary motility. Disordered motility results in ineffective mucociliary transport and accounts for bronchiectasis, sinusitis, and chronic respiratory tract infections.

        c) While females with Kartagener's syndrome may bear children, most males are sterile due to immotile spermatozoa.

        d) The condition is very similar to the immotile cilia syndrome in which patients with normal situs who have missing dynein arms suffer from chronic sinusitis, chronic bronchitis or bronchiectasis, chronic otitis media, and sterility in men.

        e) It has been postulated that normal ciliary motility is necessary for normal levorotation of the viscera during embryogenesis. Absence of the dynein arms with disordered ciliary motility would result in an equal chance of levorotation or dextrorotation. This would explain the occurrence of bronchiectasis and chronic respiratory infections in siblings of Kartagener's syndrome patients who have normal visceral situs.

      2) Cardiac anomalies occasionally occur including TGV and single ventricle.

B. Mesocardia

  1. Denotes that the heart and cardiac apex are midline in position

  2. Represents a mild form of dextrocardia

  3. Mesocardia with situs solitus

    a. Frequently has no associated cardiac abnormality

    b. L-TGV is the most common associated cardiac abnormality.

  4. Mesocardia with situs inversus

    a. May have no cardiac defect

    b. Can be associated with noncomplex cardiac abnormalities

C. Levoposition and levocardia

  1. Levoposition with situs solitus

    a. In this condition the heart is positioned further to the left than normal.

   b. There are several causes.
      1) Congenital absence of the pericardium
         a) Absence of the left side of the pericardium is more common than complete absence of the pericardium.
         b) The heart is typically markedly levopositioned on the frontal chest radiograph. On the lateral chest radiograph the cardiac shadow is rotated posteriorly. The retrosternal clear space in front of the heart is deepened because of posterior displacement of the heart. The LV may overlap with the spine.
         c) On the frontal radiograph either the MPA, the left atrial appendage (LAA), or both may be prominent because either can protrude through the pericardial defect. Prominence of the MPA is most common.
         d) CT or MR shows lung situated between either the aorta and MPA (most common) or between the MPA and LAA (normally the left pericardium keeps lung out of these recesses).
      2) CPVS — although CPVS is a predominantly right-sided condition, approximately 25% of cases occur on the left side. In these cases the heart may be abnormally levopositioned because of hypoplasia of the left lung.
   2. Levocardia with situs solitus
      a. This is the normal arrangement.
   3. Levocardia with situs inversus
      a. This is the equivalent of dextrocardia with situs solitus.
      b. The aortic knob, descending thoracic aorta, and stomach are on the right, while the heart is rotated so that its apex is on the left.
      c. There is a high incidence of associated cardiac defects including L-TGV, D-TGV, VSD, complete atrioventricular (AV) canal, pulmonic stenosis or atresia, and anomalies of systemic and pulmonary venous return.
      d. Cyanosis is usually present.

III. Visceral heterotaxy and the syndromes of asplenia and polysplenia
A. Visceral heterotaxy
   1. Refers to the abnormal arrangement of organs that differs from either situs solitus or situs inversus
   2. There is a symmetrical visceral configuration (isomerism), with duplication of either left-sided or right-sided structures.
   3. This situation is also called situs ambiguus or situs indeterminus.
   4. There is a high incidence of splenic abnormalities and major malformations of the cardiovascular system.
   5. The morphology of the atria corresponds closely with situs of the tracheobronchial tree and abdominal situs.
      a. The anatomic RA is characterized by a broad, pyramidal atrial appendage and connection to the IVC.

b. The anatomic LA has a thin atrial appendage with a narrow neck and does not connect to the IVC.

B. Asplenia (Ivemark's syndrome)
1. Bilateral right-sidedness
2. Males are affected twice as often as females.
3. The spleen, a left-sided organ, is absent.
   a. Heinz bodies and Howell-Jolly bodies are present in the erythrocytes on peripheral smear.
   b. There is a significantly increased incidence of repeated serious infections and sepsis.
4. There is visceral symmetry with abdominal heterotaxy.
   a. Bilateral anatomic RA
   b. Bilateral anatomic "right" lungs
      1) Both lungs have 3 lobes.
      2) Bilateral minor fissures
      3) Bilateral eparterial bronchi
   c. Transverse or symmetrical midline liver with right and left lobes being the same size in 50%
   d. No spleen
   e. Dextrogastria — 50%
   f. Most patients have malrotation of bowel.
   g. Genitourinary abnormalities in 15% including horseshoe kidney and hydroureter
   h. Fused or horseshoe adrenals
5. Cardiovascular abnormalities
   a. Almost all patients have multiple, severe anomalies usually of the cyanotic type.
   b. Complete AV canal occurs in almost 100% of cases with large atrial septal defect (ASD) or common atrium and high VSD.
   c. Patent ductus arteriosus (PDA) — 56%
   d. Absent coronary sinus — 85%
   e. Dextrocardia — 40%
   f. Bilateral superior vena cava (SVC) — 50%; right SVC — 34%; left SVC — 10%
   g. Totally anomalous pulmonary venous return (TAPVR) and PAPVR — TAPVR most common — 72%; PAPVR — 12%
   h. Right aortic arch common — 38%
   i. Severe pulmonic stenosis or atresia common — 80%
   j. Single ventricle — 44%
   k. D-TGV and L-TGV — 72%
   l. Rarely, double-outlet right ventricle (DORV), hypoplastic left heart, or single coronary artery
6. Radiographic findings
   a. Cardiomegaly
   b. Pulmonary vascularity is decreased in most cases because of severe pulmonic stenosis or atresia.

    c. Dextrocardia — 40%

    d. Bilateral minor fissures

    e. Bilateral eparterial bronchi

    f. Symmetrical liver occupying the entire upper abdomen — 50%

    g. Dextrogastria — 50%

    h. Malrotation of bowel

    i. Right aortic arch common — when present, catheters in the aorta and IVC lie on same side of spine

    j. Liver-spleen scan or abdominal CT demonstrates no spleen

  7. Poor prognosis

    a. High incidence of repeated episodes of infection and sepsis

    b. Severe cardiac anomalies often not readily amenable to surgical correction

C. Polysplenia

  1. Bilateral left-sidedness

  2. Slight female predominance

  3. There are multiple aggregates of splenic tissue. Added together the multiple spleens approximate the total splenic mass of a normal spleen.

  4. In polysplenia the associated anomalies tend to be fewer and less severe than in asplenia. About 25% of cases have no serious abnormality.

  5. There is visceral symmetry with abdominal heterotaxy.

    a. Bilateral anatomic LA

    b. Bilateral anatomic "left" lungs

      1) Both lungs have 2 lobes.

      2) No minor fissure on either side

      3) Bilateral hyparterial bronchi

    c. Transverse or symmetric liver common; however, the liver may be either predominantly right-sided, or left-sided.

    d. Two or more spleens

    e. Dextrogastria — common

    f. Gallbladder abnormalities including biliary atresia

    g. Malrotation of the bowel can occur.

    h. Genitourinary abnormalities

  6. Cardiovascular abnormalities

    a. Cardiovascular abnormalities are not as common and tend to be much less severe than in asplenia. Most patients are acyanotic.

    b. Azygos continuation of interrupted IVC — 85%

      1) Hepatic segment of IVC absent

      2) The azygos vein is markedly enlarged because it carries the majority of systemic venous return from the lower body.

      3) The hepatic veins still drain into the right atrium via the suprahepatic portion of the IVC. The lateral chest radiograph still demonstrates an IVC shadow in the majority of cases.

      4) In some cases, continuation of IVC flow is via the hemiazygos vein rather than the azygos vein. This is rare.

    c. ASD or AV canal — 50%

       d. VSD — 67%

       e. PDA — 50%

       f. Absent coronary sinus — 40%

       g. Dextrocardia — 40%

       h. Bilateral SVC — 33%; right SVC — 33%; left SVC — 33%

       i. PAPVR and TAPVR (50%) — PAPVR most common, usually with each lung returning blood to its ipsilateral atrium

       j. Right aortic arch — 67%

       k. Abnormalities of the pulmonic valve are much less common.

          1) The pulmonic valve is normal in 60%.

          2) 33% have mild pulmonic stenosis

          3) Pulmonic atresia is rare.

       l. Single ventricle, D-TGV, L-TGV, DORV, and other serious cardiac abnormalities are uncommon.

    7. Radiographic findings

       a. Cardiomegaly

       b. Pulmonary vascularity is increased in 85% usually due to associated L-R shunt.

       c. Enlarged azygos vein secondary to azygos continuation of an interrupted IVC — 85% (this is difficult to appreciated from plain films in infants and small children but is much more obvious in adolescents and adults with polysplenia)

       d. Dextrocardia — 40%

       e. No minor fissure

       f. Bilateral hyparterial bronchi

       g. Transverse or symmetric liver common; however, the liver may be either predominantly right-sided, or left-sided.

       h. Dextrogastria is common — if the aortic arch and gastric air bubble are on opposite sides (discordant), there is a uniform tendency for congenital interruption of IVC with azygos continuation

       i. Malrotation of bowel

       j. Liver-spleen scan or abdominal CT demonstrates multiple spleens.

          1) These may all be located in the left upper quadrant.

          2) In some cases the multiple spleens are spread throughout the mesentery.

    8. Better prognosis

       a. A significant number of cases of polysplenia have azygos continuation of the IVC as the only associated abnormality.

       b. Milder cardiac anomalies often amenable to surgery

       c. No increased incidence of infection or sepsis

IV. Secondary dextrocardia or levocardia

    A. Refers to abnormal positioning of the heart secondary to mediastinal shift from pulmonary or thoracic cage abnormalities

       1. Scoliosis

       2. Sternal or rib deformity

John H. Woodring, M.D.

3. Pneumonectomy or lobectomy
4. Lobar or whole lung collapse
5. Pneumothorax
6. Massive pleural effusion
7. Unilateral air-trapping
8. Diaphragmatic hernia

References and suggested additional reading

1. Fuster V, Alexander RW, O'Rourke RA, Roberts R, King SB III, Wellens HJJ. *Hurst's The Heart 10th ed.* New York: McGraw-Hill, 2001.
2. Gedgaudas E, Moller JH, Castaneda-Zuniga WR, Amplatz K. *Cardiovascular Radiology*. Philadelphia: Saunders, 1985.
3. Miller SW. *Cardiac Radiology*: The Requisites. St. Louis: Mosby, 1996.
4. Spindola-Franco H, Fish BG. *Radiology of the Heart: Cardiac Imaging in Infants, Children, and Adults*. New York: Springer-Verlag, 1985.
5. Woodring JH, Howard TA, Kanga JF. Congenital pulmonary venolobar syndrome revisited. RadioGraphics 1994; 14:349-369.
6. Woodring JH, Royer JM, McDonagh D. Kartagener's syndrome. JAMA 1982; 247:2814-2816.

Chapter 10.

Congenital anomalies of the pulmonary arteries and veins

I. **Congenital anomalies of the pulmonary arteries**
 A. **Embryologic considerations**
  1. By the third week of fetal development there are two primitive ventral (ascending) aortae and two primitive dorsal (descending) aortae.
  2. During the fourth week the paired ventral aortae fuse to form a single midline trunk — the aortic sac — that connects to the short truncus of the heart. The paired dorsal aortae fuse to form a midline descending aorta, and the first two pairs of symmetrical branchial aortic arches develop between the ventral and dorsal aortae.
  3. Six pairs of branchial aortic arches and paired seventh intersegmental arteries arising from the proximal descending aorta develop during the fourth week and their transformation mainly occupies the fifth, sixth, and seventh weeks of fetal development.
   a. The first and second paired aortic arches disappear as soon as the third and fourth arches develop.
   b. The third arches eventually become the internal carotid arteries.
   c. The fourth arches form the right and left embryologic aortic arches.
   d. The fifth aortic arches form and quickly regress, as do the portions of the primitive dorsal aortae between the third and fourth arches.
   e. The sixth paired aortic arches persist and eventually become the pulmonary arteries and ductus arteriosi.
   f. The seventh intersegmental arteries migrate superiorly and become the subclavian arteries.
  4. As the branchial arches are forming and regressing, changes are occurring in the aortic sac and truncus. A truncoconal (spiral) septum forms and separates the aortic sac into two halves. The left side or pulmonary portion becomes the main pulmonary artery (MPA) and the right side or aortic portion becomes the ascending aorta.
  5. At the same time the ostia of the sixth aortic arches migrate to the left to attach to the MPA. The proximal portions of the sixth arches become the proximal right and left pulmonary arteries while the distal sixth aortic arches become the right and left ductus arteriosi. Outpouchings from the proximal portions of the sixth aortic arches penetrate the developing lung buds to form the pulmonary arterial tree within the lungs.
  6. Following this the dorsal segment of the embryonic right aortic arch between the right subclavian artery and the dorsal aorta involutes and the right ductus arteriosus regresses. The anterior remnant of the right aortic arch becomes the innominate artery and the descending aorta shifts to the left. This results in the normal arrangement of a left aortic arch, which

has as its branches, an innominate artery followed by the left common carotid and left subclavian arteries.

7. Many of the congenital anomalies of the pulmonary arteries can be explained on the basis of faulty septation of the aortic sac and truncus or abnormal development of the sixth branchial aortic arches.

B. **Supravalvular stenosis of the MPA and peripheral pulmonic stenosis**
   1. Can occur as an isolated anomaly or as part of William's syndrome
   2. There are four types.
      a. Type I — stenosis of the MPA
      b. Type II — stenosis at the bifurcation of the MPA which causes narrowing of the distal MPA and proximal right and left pulmonary arteries
      c. Type III — multiple peripheral stenoses with normal MPA and normal central right and left pulmonary arteries
      d. Type IV — central and peripheral stenoses often resulting in diffuse hypoplasia of the MPA and central pulmonary arteries with associated peripheral stenoses
   3. The chest radiographic appearance is variable depending upon which type of supravalvular pulmonic stenosis the patient has.
   4. The diagnosis is usually made by pulmonary arteriography although MR can also confirm the diagnosis.

C. **Congenital interruption of the proximal right or left pulmonary artery**
   1. Is due to interruption of the proximal portion of the primitive right or left sixth aortic arch
   2. Congenital interruption of the proximal left pulmonary artery (LPA) usually occurs in association with tetralogy of Fallot (TOF).
      a. The left hilum and pulmonary vessels in the left lung are quite small and there is systemic arterial collateral flow to the left lung.
      b. The bronchial arteries and intercostal arteries serve as the main sources of collateral blood supply to the left lung.
      c. There may be associated unilateral inferior rib notching on the left side.
   3. Congenital interruption of the proximal right pulmonary artery (RPA) usually occurs in association with hypoplasia of the right lung. Usually there are no other cardiac abnormalities, although there is a slight increase in the incidence of acyanotic left-to-right (L-R) shunts and congenital abnormalities of the aortic arch.
      a. The bronchial arteries and intercostal arteries serve as the main sources of collateral blood supply to the right lung.
      b. Radiographic findings
         1) The right lung is usually hypoplastic and smaller than the left, the right hilum is small, and the right lung is often of decreased density due to oligemia.
         2) The left hilum and pulmonary vessels in the left lung are enlarged since the left lung receives the entire right ventricular output.

3) Bronchial arterial collaterals may be seen as serpiginous or "cork-screw" vessels near the right hilum. Transpleural collaterals from the intercostal arteries are more pronounced in the lung periphery and have a serpiginous almost "reticulonodular" appearance and can even be somewhat linear in configuration resembling Kerley A or B lines.

4) Unilateral inferior rib notching on the right side may also occur.

c. Most patients are asymptomatic. In some cases pulmonary arterial hypertension and congestive heart failure develop due to the chronic overcirculation of the left lung. Patients with interruption of the proximal RPA may also experience hemoptysis due to bleeding from the systemic arterial collateral arteries.

D. Truncus arteriosus

1. Results from failure of formation of the truncoconal (spiral) septum within the truncus in the early embryo
2. As a result, a single arterial vessel (the truncus) leaves the heart.
3. The pulmonary arteries arise from the truncus.
4. There are three types.
   a. Type I (60%)
      1) Most common form
      2) A short MPA arises from the left side of the truncus and divides into right and left pulmonary arteries.
      3) Because the MPA is anatomically present, the chest radiograph frequently shows an enlarged MPA.
      4) There is moderate cardiomegaly with enlargement of both ventricles and the left atrium (LA) associated with overcirculation of the pulmonary vascularity.
   b. Type II (30%)
      1) Each pulmonary artery arises independently from the posterolateral aspect of the left side of the truncus.
      2) There is no MPA; the MPA segment of the left heart border is concave.
      3) There is moderate cardiomegaly with enlargement of both ventricles and the LA associated with overcirculation of the pulmonary vascularity.
   c. Type III (10%)
      1) The RPA arises from the right lateral wall of the truncus; the LPA arises from the left lateral wall of the truncus.
      2) There is no MPA; the MPA segment of the left heart border is concave.
      3) There is moderate cardiomegaly with enlargement of both ventricles and the LA associated with overcirculation of the pulmonary vascularity.
5. Mirror-image right aortic arch occurs in 40%.

E. Aorticopulmonary (AP) window

1. Rare abnormality

2. Results from faulty formation of the truncoconal (spiral) septum that leaves a large communication between the ascending aorta and main pulmonary artery.

3. There are three forms
   a. Type I — proximal communication between ascending aorta and MPA above the origin of the coronary arteries
   b. Type II — distal communication between the posterior aspect of the ascending aorta and MPA near the origin of the right pulmonary artery
   c. Type III — complete defect with communication of the entire ascending aorta and MPA from the level of the valves to the origin of the right pulmonary artery

4. Usually manifested by cardiomegaly and marked overcirculation of the pulmonary vascularity. The MPA is usually markedly enlarged.

F. **Anomalous origin of one pulmonary artery from the ascending aorta (hemitruncus)**
   1. Rare anomaly probably related to faulty septation of the truncus causing one of the sixth branchial aortic arches to become incorporated into the ascending aorta.
   2. Flow patterns in the affected fetus are thought to favor persistence of the contralateral fourth branchial aortic arch, therefore, the aortic arch is found on the side opposite the ectopic pulmonary artery.
   3. Respiratory distress, cyanosis, and congestive heart failure present in infancy.
   4. Anomalous RPA
      a. More common type
      b. Left aortic arch
      c. May be associated with patent ductus arteriosus (PDA) or occasionally AP window
      d. Usually no ventricular septal defect (VSD)
      e. Chest radiographs typically show unilateral overcirculation of the R lung.
   5. Anomalous LPA
      a. Less common type
      b. Usually occurs as part of TOF
      c. Right aortic arch
      d. May have absent pulmonic valve
      e. VSD common, rarely may have persistent right-sided PDA
      f. Chest radiographs demonstrate unilateral overcirculation of the left lung in a patient with a right aortic arch.
      g. May have lobar emphysema on the right from compression of the right bronchial tree between the right aortic arch or right-sided PDA and the RPA

G. **Pulmonary artery sling (aberrant LPA)**
   1. Rare abnormality
   2. The LPA arises from the proximal portion of the RPA.

3. The aberrant LPA courses through the mediastinum between the trachea and esophagus at the level of the right hilum to reach the left lung.
4. Clinical manifestations
   a. Stridor on expiration — with other types of vascular rings stridor is usually more pronounced on inspiration
   b. Dysphagia uncommon
5. Usually an isolated anomaly
6. May have associated defects including:
   a. "Ring-sling" complex (54%)
      1) The tracheal cartilages in the distal trachea and mainstem bronchi form complete rings.
      2) Associated with significant stenosis of the distal trachea and mainstem bronchi
      3) Early onset of stridor and other respiratory symptoms
      4) Stenosis of the trachea and mainstem bronchi may persist after surgical relocation of the aberrant LPA.
   b. Atrial septal defect (ASD)
   c. VSD
   d. PDA
   e. TOF
   f. Single ventricle
   g. Persistent left superior vena cava (SVC)
7. Imaging findings
   a. The posteroanterior chest radiograph demonstrates a round convexity of the upper right hilar shadow that represents the origin of the aberrant LPA.
   b. The aberrant LPA causes compression and narrowing of the distal trachea and right mainstem bronchus.
      1) There may be opacification of the right upper lobe from retained fetal lung fluid.
      2) After the retained fetal lung fluid clears, there may be postobstructive hyperinflation of the right upper lobe or right lung.
   c. The lateral chest radiograph shows a posterior indentation on the lower trachea at the level of the right hilum.
   d. Barium swallow demonstrates an anterior indentation on the esophagus at the level of the right hilum with a rounded soft tissue density between the distal trachea and esophagus that represents the aberrant LPA.
   e. The diagnosis can be confirmed by pulmonary arteriography, contrast-enhanced CT, or MR. CT and MR nicely demonstrate the origin of the LPA from the RPA and its course between the trachea and esophagus to reach the left lung.
   f. In those with the ring-sling complex, the narrowed distal trachea and main bronchi have the configuration of an inverted "T."
8. Pulmonary artery sling is the most notable vascular ring to pass between the trachea and esophagus. Ductus arteriosus sling and occasionally

aberrant right subclavian artery also go between the trachea and esophagus. All other vascular rings are posterior and lateral to both the trachea and esophagus.

H. Ductus arteriosus sling
1. Very rare abnormality
2. The ductus arteriosus arises from the RPA and courses through the mediastinum between the trachea and esophagus to reach the aortic isthmus.
3. Clinical symptoms similar to pulmonary artery sling
4. Plain film findings similar to pulmonary artery sling
5. Aortography demonstrates a patent ductus that communicates with the RPA.
6. May be associated with aberrant right subclavian artery

I. Pulmonary arteriovenous malformation (AVM) and arteriovenous (AV) fistula
1. Pulmonary AVM refers to a spectrum of vascular communications between the pulmonary arteries and pulmonary veins ranging from small microscopic communications too small to identify by imaging means, to macroscopic peripheral communications between the pulmonary arteries and veins fed by a single pulmonary artery and drained by a single pulmonary vein (simple pulmonary AVM), to large complex communications between the central pulmonary arteries and veins at the hilar level (pulmonary AV fistula) that may have one or more feeding arteries and draining veins.
2. 80-90% of AVMs consist of a single feeding artery and single draining vein (simple pulmonary AVM)
3. Pulmonary AVMs are solitary in 66% of cases and are multiple in 33%.
4. 70% of patients with pulmonary AVMs have abnormal arteriovenous communications in the skin, mucous membranes, and other organs. This disorder, known as Osler-Weber-Rendu disease or hereditary hemorrhagic telangiectasia, has an autosomal dominant non-sex-linked inheritance.
5. Although the AVM is probably present at birth, 90% are diagnosed in adults.
6. Clinical signs and symptoms
   a. Hemoptysis is the most common clinical manifestation.
   b. Pulmonary AVM produces a right-to-left (R-L) shunt that allows unoxygenated blood to bypass the pulmonary capillary bed.
      1) Dyspnea occurs in 60% of cases.
      2) Cyanosis and finger clubbing may be present if the AVMs are large or numerous.
   c. Central nervous system symptoms may result from metastatic abscess, hypoxemia, cerebral thromboembolism, and cerebral hemorrhage from concomitant cerebral AVM.
7. Imaging findings
   a. Simple AVM

1) Lobular pulmonary nodule with feeding artery and draining vein. The draining vein is usually larger than the feeding artery.
2) The diagnosis can often be made from plain films alone. CT and MR are both good at confirming the diagnosis. Pulmonary arteriography is still the gold standard and is usually done prior to treatment.
   b. AV fistula
      1) Typically seen as a lobular hilar mass that can be quite large
      2) Contrast-enhanced CT or MR can establish the vascular nature of the lesion.
   c. Chest wall involvement
      1) Occasionally an AVM may receive part of its blood supply from intercostal arteries in the adjacent chest wall.
      2) When this happens the patient may develop unilateral inferior rib notching on the ipsilateral side.
8. Treatment
   a. Pulmonary AVMs can be surgically resected or occluded by vascular coils.
   b. In some cases, resection or coil occlusion of pulmonary AVMs results in previously unrecognized microscopic AVMs enlarging and becoming macroscopic.
   c. Pulmonary AV fistulas usually require surgical closure.

II. Congenital anomalies of the pulmonary veins
A. Embryologic considerations
1. Normally, the pulmonary veins arise from a capillary plexus, derived from the general splanchnic plexus, which communicates freely with both the cardinal and umbilicovitelline venous systems.
2. At 27 to 29 days of fetal development, the common pulmonary vein is seen as an outpouching from the superior wall of the LA, growing toward the lung bud.
3. By 28 to 30 days of fetal development, the common pulmonary vein communicates with the pulmonary venous plexus, which is becoming separated from the main splanchnic plexus.
4. By 30 to 32 days, the pulmonary venous plexus loses most of its connections with the splanchnic plexus, and blood flows from the pulmonary venous plexus through the common pulmonary vein to the LA; however, small communications with derivatives of the cardinal and umbilicovitelline venous system persist. At about the same time, the common pulmonary vein becomes incorporated into the posterior LA wall, and thereby the four pulmonary veins come to enter separately into this chamber.
B. Totally anomalous pulmonary venous return (TAPVR)
1. TAPVR occurs when the common pulmonary vein fails to become incorporated into the LA and remains as a separate chamber from the LA. The pulmonary venous system maintains connections with derivatives of

the right or left cardinal venous system, or umbilicovitelline venous system.

2. An ASD is an integral part of the anomaly and allows blood to flow from the right atrium (RA) to the LA.

3. There are four types.

    a. Supracardiac connection — 55%

        1) Connection to a right superior vena cava (RSVC) (10%)

            a) Represents persistence of communication to the right anterior cardinal system (supracardiac connection)

            b) There is overcirculation of the pulmonary vascularity.

            c) The MPA is often enlarged.

            d) The upper right margin of the mediastinum is prominent and convex outwards due to enlargement of the SVC that is carrying both systemic venous return from the upper body and pulmonary venous return from both lungs.

            e) The cardiac silhouette is enlarged with both RAE and RVE.

            f) The ASD allows R-L shunting of admixed blood to the LA; however, the LA is not enlarged.

        2) Connection to a left superior vena cava (LSVC) or "vertical vein" (45%)

            a) Represents persistence of communication to the distal part of the left anterior cardinal system (supracardiac connection)

            b) There is overcirculation of the pulmonary vascularity.

            c) There is a characteristic "snowman" or "figure-of-8" configuration of the heart and upper mediastinum due to a convex outward bulge in the upper left mediastinal border from the prominent LSVC or vertical vein and a convex outward bulge in the upper right mediastinal border from the enlarged RSVC. Blood flows from the common pulmonary vein up the LSVC or vertical vein, crosses the mediastinum via the left brachiocephalic or innominate vein, and enters the RSVC that is also enlarged. These enlarged veins produce the upper half of the "snowman" and the heart forms the lower half. These enlarged veins also produce a pretracheal soft-tissue "mass effect" on the lateral chest radiograph.

            d) The MPA is enlarged but this may be obscured on the plain film by the upper half of the "snowman."

            e) The cardiac silhouette is enlarged with both RAE and RVE.

            f) The ASD allows R-L shunting of admixed blood to the LA; however, the LA is not enlarged.

    b. Cardiac connection — 30%

        1) Represents persistence of communication to the proximal part of the left anterior cardinal system (cardiac connection)

        2) Connection to the coronary sinus (16%)

        3) Connection to RA (14%)

4) There is no characteristic cardiac configuration. The radiographic appearance is identical to ASD.
   a) Overcirculation of pulmonary vascularity
   b) Enlarged MPA
   c) Cardiomegaly with RAE, RVE, and no LAE

c. Connection to the portal venous system with obstruction — 12%
   1) Represents persistence of communication to the umbilical-vitelline system (infracardiac connection)
   2) The anomalous right and left pulmonary veins converge into a single venous trunk that follows the esophagus through the esophageal hiatus in the diaphragm into the abdomen to connect to the portal venous system. Pulmonary venous return is typically obstructed at the level of the diaphragm where the anomalous pulmonary vein courses through the esophageal hiatus.
   3) Characterized by severe reticulation
   4) The heart size is normal. The MPA is not prominent.

d. Mixed lesions — 3%

C. Partially anomalous pulmonary venous return (PAPVR)
1. PAPVR occurs when there is incomplete fusion of the common pulmonary vein with the pulmonary venous plexus. When this occurs, one or more of the pulmonary veins become incorporated into the LA, while the others retain their communications with the cardinal or umbilicovitelline venous systems and drain to the RA or one of its tributaries.
2. There are two forms of PAPVR.
   a. PAPVR associated with sinus venosus type ASD
   b. PAPVR associated with congenital pulmonary venolobar syndrome
3. Sinus venosus type ASD
   a. 6% of cases of ASD
   b. Almost always associated with PAPVR from the right upper lobe
   c. The right superior pulmonary vein does not cross the descending branch of the right pulmonary artery to enter the LA as it normally does. Instead, the right superior pulmonary vein enters either the SVC or RA.
4. Congenital pulmonary venolobar syndrome (CPVS)
   a. CPVS represents a group of congenital abnormalities of the thorax that may occur alone or together in almost any combination. It was given the name venolobar syndrome because of the frequent occurrence of either an abnormal lobe of one of the lungs or an anomalous pulmonary vein.
   b. Major or common components of CPVS
      1) Hypogenetic lung
         a) Hypogenetic lung is a term used to encompass three entities: pulmonary agenesis, pulmonary aplasia, and pulmonary hypoplasia.

b) In pulmonary agenesis there is complete absence of the lobe and its bronchus.

c) In pulmonary aplasia there is no lung tissue present; however, there is a rudimentary bronchus that ends in a blind pouch.

d) In pulmonary hypoplasia the alveoli and bronchi are present but the involved lobe is small.

e) Hypogenetic lung is the most common component of CPVS and occurs in 69% of cases.

f) Hypogenetic lung is much more common on the right, with 75% of cases involving the right lung and only 25% involving the left lung.

2) PAPVR

a) In CPVS part or all of the hypogenetic lung may be drained by one or more anomalous pulmonary veins.

b) The anomalous vein usually drains into the inferior vena cava (IVC) below the right hemidiaphragm; less commonly PAPVR drains into the suprahepatic IVC, hepatic veins, portal vein, azygos vein, coronary sinus, or RA.

c) PAPVR results in a L-R shunt which is usually asymptomatic unless the shunt is 2:1 or greater.

d) The curvilinear configuration of the anomalous vein as it courses toward the right hemidiaphragm has been likened to a Turkish sword or scimitar. Thus, when hypogenetic lung and PAPVR occur together the condition is referred to as scimitar syndrome.

e) PAPVR is the second most common component of CPVS occurring in 31% of cases of CPVS and 40% of those with hypogenetic lung.

3) Congenital interruption of the proximal RPA

a) Congenital interruption of the proximal RPA occurs in 14% of cases of CPVS.

b) Almost all cases are associated with hypoplasia of the right lung.

c) Systemic arterial collaterals are frequently visible on plain films.

d) Unilateral inferior rib notching of the right ribs may occur.

4) Pulmonary sequestration

a) Two main forms

i) Intralobar sequestration

ii) Extralobar sequestration

iii) See Chapter 8 for detailed discussion of pulmonary sequestration

b) Occurs in 24% of cases of CPVS.

c) May be associated with hypogenetic lung and systemic arterialization of the lung without sequestration

5) Systemic arterialization of the lung without sequestration

a) Systemic arterialization of the lung without sequestration occurs on the right side in approximately 10% of cases of CPVS.

b) Typically the right lung is hypogenetic and there may be PAPVR (scimitar syndrome), accessory right hemidiaphragm, and other features of CPVS.

c) Part or all of the hypogenetic right lung may receive its arterial blood supply from the thoracic or abdominal aorta or their branches.

d) Systemic arterialization of the lung without sequestration also occurs in horseshoe lung, which is a rare minor or uncommon component of CPVS (see below).

6) Azygos continuation of an interrupted IVC

a) Azygos continuation of an interrupted IVC may occur as an isolated anomaly, as part of the polysplenia syndrome, or as part of CPVS.

b) Azygos continuation of an interrupted IVC occurs in approximately 17% of cases of CPVS.

c) Although the literature clearly documents that the shadow of the IVC may be missing on the lateral chest radiograph, the hepatic veins usually still drain to the suprahepatic portion of the IVC and the lateral chest radiograph usually still shows an IVC shadow.

7) Accessory (or duplicated) right hemidiaphragm

a) Accessory right hemidiaphragm is one of the less common major components of CPVS, occurring in only about 7% of cases.

b) An accessory right hemidiaphragm is a thin membrane in the right hemithorax that is fused anteriorly with the diaphragm and that courses posterosuperiorly to join the posterior chest wall. The accessory right hemidiaphragm thus separates the right hemithorax into two parts, trapping part or all of the right middle or lower lobes beneath it. The pulmonary vessels and bronchi must pass through the accessory right hemidiaphragm to get to the lung trapped below it.

c) If the lung beneath an accessory right hemidiaphragm is not aerated it will appear as a solid mass along the posterior aspect of the right hemidiaphragm. If the lung beneath the accessory right hemidiaphragm is aerated, the accessory right hemidiaphragm will appear as a fissure-like oblique line in the posterior lower right hemithorax on the lateral view. The line courses from the anterior aspect of the diaphragm to the posterior chest wall.

d) Accessory right hemidiaphragm usually occurs in patients with hypogenetic lung. In those who also have PAPVR (scimitar syndrome) the anomalous pulmonary vein must travel through the accessory right hemidiaphragm and then the main right hemidiaphragm to reach the systemic venous system.

  c. Minor or uncommon components of CPVS
    1) Tracheal trifurcation
      a) Extremely rare component of CPVS
      b) The trachea divides into three main bronchi with two supplying the right lung and one supplying the left.
    2) Diaphragmatic abnormalities other than accessory right hemidiaphragm.
      a) Occasionally patients with CPVS may demonstrate diaphragmatic abnormalities other than accessory right hemidiaphragm.
      b) These include congenital weakness or eventration of the diaphragm, partial absence of the diaphragm, and phrenic cyst.
    3) Horseshoe lung
      a) This is a rare congenital malformation that occurs in association with other components of CPVS including hypogenetic lung, PAPVR, and systemic arterialization of the lung without sequestration.
      b) The right and left lungs are fused posteriorly behind the heart by an isthmus of pulmonary parenchyma arising from the right lung base.
      c) The isthmus may be fused to the left lung without intervening pleura or with intervening pleura that resembles an accessory fissure. When intervening pleura is present, the frontal chest radiograph shows a characteristic linear or curvilinear fissure at the left lung base behind the heart.
      d) The diagnosis can be made by pulmonary arteriography that shows a branch arising from the inferior aspect of the proximal RPA crossing behind the heart to supply part of the left lung. The right lower lobe bronchus also gives off an accessory bronchus that crosses the midline through the isthmus to supply part of the left lung.
      e) CT can confirm the diagnosis by demonstrating the isthmus of lung connecting the right and left lungs through a defect in the mediastinum behind the heart.
      f) Unlike patients with simple scimitar syndrome, there is a high prevalence of associated pulmonary infection in patients with horseshoe lung.
    4) Esophageal and gastric lung
      a) These are rare variants of pulmonary sequestration in which the bronchus of the sequestered lung tissue communicates with either the esophagus or stomach.
      b) Sequestration may be of either the intralobar or extralobar type.
      c) Patients are typically seen in infancy with coughing on feeding and recurrent pulmonary infections.

5) Aberrant course of the SVC
   a) In most cases of scimitar syndrome the scimitar shadow is cast by an anomalous pulmonary vein.
   b) Rarely, the scimitar shadow may be cast by the SVC that follows an aberrant course through the substance of the right lung to reach the RA.
6) Absence of the left pericardium
   a) Usually occurs as an isolated anomaly
   b) Can rarely occur in CPVS

  d. Other associated anomalies
    1) Cardiac abnormalities occur in 25-50% of cases.
     a) ASD is most common.
     b) VSD
     c) PDA
     d) Atrioventricular canal
     e) Coarctation of aorta
     f) Hypoplastic left heart syndrome
     g) TOF
     h) Double-outlet right ventricle
     i) Pulmonary arterial hypertension
    2) Spinal abnormalities occasionally occur.
     a) Scoliosis
     b) Hemivertebra
    3) Rare associated anomalies
     a) Mental retardation
     b) Omphalocele
     c) Abnormalities of the genitourinary tract, peripheral skeleton, and eyes.

D. Meandering pulmonary veins
1. Meandering pulmonary veins probably represent a variant or incomplete form of PAPVR in which delayed fusion of the common pulmonary vein with the pulmonary venous plexus causes the pulmonary veins to follow an anomalous course through the lungs to drain to the LA.
  a. When the abnormality involves the inferior pulmonary veins, the aberrant veins descend toward the diaphragm as if they were going to communicate with derivatives of the umbilicovitelline venous system. However, as the aberrant inferior pulmonary veins come near the diaphragm they turn upward to reach the LA. Thus the shadows cast by the aberrant or "meandering" inferior pulmonary veins resemble a scimitar shadow except that the veins can be followed to the LA rather than to the diaphragm.
  b. The course of the superior pulmonary veins may be abnormal as well. Normally the superior pulmonary veins cross in front of the right and left bronchi to enter the LA. In meandering pulmonary veins the superior pulmonary veins may cross behind the bronchi to enter the LA.

John H. Woodring, M.D.

2. Meandering pulmonary veins can be an isolated anomaly, but they are often associated with hypogenesis of the right lung and dextropositioning of the heart.

E. Cor triatriatum
1. Represents incomplete incorporation of the common pulmonary vein into the LA.
2. The common pulmonary vein persists as a separate chamber that communicates with the LA through a small opening which is usually stenotic.
3. Radiographically, cor triatriatum is characterized by pulmonary venous hypertension with severe reticulation. The MPA is prominent and there may be RVE.

F. Stenosis or atresia of the pulmonary veins
1. May be congenital or acquired
2. Acquired causes include fibrosing mediastinitis, tumor invasion, and constrictive pericarditis.
3. Congenital stenosis or atresia of the pulmonary veins probably results from faulty fusion of the common pulmonary vein with the pulmonary venous plexus resulting in stenosis or atresia of one or more of the pulmonary veins as they enter the common pulmonary vein.
4. Radiographic findings include:
   a. Severe reticulation which can be unilateral if the stenosis is unilateral
   b. The MPA segment may be prominent and there may be RVE.

G. Pulmonary venous varix
1. Can be either congenital or acquired
   a. The acquired form is more common than the congenital form.
   b. Acquired pulmonary venous varix is usually associated with long-standing mitral regurgitation and is thought to be due to a jet of regurgitant blood from the mitral valve that selectively dilates one of the pulmonary veins. If a patient with a pulmonary venous varix has no evidence of mitral valve disease then the pulmonary venous varix is probably congenital.
2. One of the pulmonary veins, typically one of the inferior pulmonary veins, is markedly enlarged and varicose.
3. Chest radiographs typically show a lobular density behind the heart closely related to the LA. CT can confirm the diagnosis by showing the lobular enlarged pulmonary vein entering the LA.
4. Can be distinguished from AVM or AV fistula by the fact that it is not attached in any way to the pulmonary artery

References and suggested additional reading

1. Arey LB. *Developmental Anatomy a Textbook and Laboratory Manual of Embryology 7th ed.* Philadelphia: Saunders, 1965.

2. Folger GM Jr. The scimitar syndrome: anatomic, physiologic, developmental, and therapeutic considerations. Angiology 1976; 23:373-407.
3. Fuster V, Alexander RW, O'Rourke RA, Roberts R, King SB III, Wellens HJJ. *Hurst's The Heart 10th ed.* New York: McGraw-Hill, 2001.
4. Gedgaudas E, Moller JH, Castaneda-Zuniga WR, Amplatz K. *Cardiovascular Radiology*. Philadelphia: Saunders, 1985.
5. Goodman LR, Jamshidi A, Hipona FA. Meandering right pulmonary vein simulating the scimitar syndrome. Chest 1972; 62: 510-512.
6. Kriss VM, Woodring JH, Cottrill CM. "Meandering" pulmonary veins: report of a case in an asymptomatic 12-year-old girl. J Thorac Imaging 1995; 10:142-145.
7. Meszaros WT. *Cardiac Roentgenology: Plain Films and Angiocardiographic Findings*. Springfield: Charles C Thomas, 1969.
8. Miller SW. *Cardiac Radiology: The Requisites*. St. Louis: Mosby, 1996.
9. Shuford WH, Sybers RG. *The Aortic Arch and its Malformations With Emphasis on the Angiographic Features*. Springfield: Charles C Thomas, 1974.
10. Müller NL, Fraser RS, Colman NC, Paré PD. *Radiologic Diagnosis of Diseases of the Chest*. Philadelphia: Saunders, 2001.
11. Partridge JB, Osborne JM, Slaughter RE. Scimitar et cetera: the dysmorphic right lung. Clin Radiol 1988; 39:11-19.
12. Spindola-Franco H, Fish BG. *Radiology of the Heart: Cardiac Imaging in Infants, Children, and Adults*. New York: Springer-Verlag, 1985.
13. Woodring JH, Howard TA, Kanga JF. Congenital pulmonary venolobar syndrome revisited. RadioGraphics 1994; 14:349-369.

Chapter 11.

Congenital variants and anomalies of the aortic arch and great veins

I. **"Raider's triangle" and the radiographic evaluation of abnormalities of the aortic arch**
   A. **Definition of Raider's triangle**
      1. Raider's triangle is the triangular lucent area on the lateral chest radiograph that is bounded anteriorly by the posterior wall of the trachea, inferiorly by the top of the aortic arch, and posteriorly by the anterior margin of the thoracic spine.
         a. It is formed by the contact of the right and left lungs behind the esophagus.
         b. On the posteroanterior (PA) chest radiograph contact of the right and left lungs behind the esophagus forms the posterior junction line — a vertical white line seen through the trachea extending from the thoracic inlet to the top of the aortic arch.
      2. It is named in honor of Louis Raider who first highlighted its importance.
   B. **Differential diagnosis of masses in Raider's triangle**
      1. Congenital abnormalities of the aortic arch
         a. Many congenital abnormalities of the aortic arch pass through Raider's triangle and produce a soft tissue mass-like density posterior to the trachea and esophagus that may cause a posterior indentation on the trachea and esophagus and may displace the trachea and esophagus anteriorly. These include:
            1) Aberrant right subclavian artery (RSCA) arising from a left aortic arch
            2) Aortic diverticulum arising from either a left or right aortic arch
            3) Aberrant left subclavian artery (LSCA) arising from a right aortic arch
            4) Aberrant left innominate artery arising from a right aortic arch
            5) Double aortic arch
            6) Right cervical aortic arch
            7) Left aortic arch with right descending aorta
         b. Because of this, close inspection of Raider's triangle is crucial in evaluation of the lateral chest radiograph in patients with suspected congenital abnormalities of the aortic arch.
      2. True and false aortic aneurysms
      3. Esophageal abnormalities
         a. Benign and malignant tumors
         b. Congenital atresia

      c. Duplication cysts
      d. Diverticulum
      e. Achalasia

   4. Other mediastinal abnormalities can occasionally present in Raider's triangle
      a. Intrathoracic goiter
      b. Bronchogenic cyst
      c. Bronchogenic carcinoma

II. Congenital variants and anomalies of the aortic arch and its major branches

A. Embryologic considerations

   1. By the third week of fetal development there are two primitive ventral (ascending) aortae and two primitive dorsal (descending) aortae.

   2. During the fourth week the paired ventral aortae fuse to form a single midline trunk — the aortic sac — that connects to the short truncus of the heart. The paired dorsal aortae fuse to form a midline descending aorta, and the first two pairs of symmetrical branchial aortic arches develop between the ventral and dorsal aortae. Ultimately, six pairs of branchial aortic arches and paired seventh intersegmental arteries arising from the proximal descending aorta develop during the fourth week and their transformation mainly occupies the fifth, sixth, and seventh weeks of fetal development.
      a. The first and second paired aortic arches disappear as soon as the third and fourth arches develop.
      b. The third arches eventually become the internal carotid arteries.
      c. The fourth arches form the right and left embryologic aortic arches.
      d. The fifth aortic arches form and quickly regress, as do the portions of the primitive dorsal aortae between the third and fourth arches.
      e. The sixth paired aortic arches persist and eventually become the pulmonary arteries and ductus arteriosi.
      f. The seventh intersegmental arteries migrate superiorly and become the subclavian arteries.

   3. The result is an embryologic double aortic arch system.
      a. There is an aortic arch and ductus arteriosus on each side.
      b. Each aortic arch gives rise to a common carotid artery and a subclavian artery.

   4. Interruption of this embryologic double aortic arch system at different locations can explain the development of the normal aortic arch and the aortic arch anomalies.

B. Left aortic arch — 5 types

   1. Normal left aortic arch
   2. Left aortic arch with aberrant RSCA
   3. Left aortic arch with aberrant right innominate artery
   4. Left aortic arch with right ductus arteriosus
   5. Left aortic arch with right descending aorta

C. Normal left aortic arch

1. There is interruption of the dorsal segment of the right aortic arch between the RSCA and the descending aorta with regression of the right ductus arteriosus. The anterior remnant of the right aortic arch becomes the innominate artery and the descending aorta shifts to the left. This results in a left aortic arch.

2. Normal branching of a left aortic arch
    a. Right-sided innominate artery, left common carotid artery (LCCA), LSCA — most common arrangement occurring in about 70% of normals
    b. In 25% the innominate and left common carotid arise from a single trunk.
        1) This is referred to as a "bovine" aortic arch because it is the most common arrangement in cows.
        2) This is the most common anatomical variant of the aortic arch.
            a) A subtle distinction is made between an anatomical variant and a congenital anomaly.
            b) An anatomical variant is not abnormal and it produces no symptoms.
            c) A congenital anomaly is abnormal and because of this it may produce symptoms.
            d) For example, an os trigonum, a small accessory ossicle posterior to the talus, is an anatomical variant. Clubfoot is a congenital anomaly.
    c. In 5-6% the left vertebral artery arises from the aortic arch rather than the LSCA.
    d. Rarely, the LCCA and right common carotid artery (RCCA) may both arise from a single branch coming off the aortic arch (bicarotid trunk).
    e. A left-sided ductus arteriosus, and subsequently a left-sided ligamentum arteriosum, connects the inferior aspect of the distal aortic arch to the left pulmonary artery (LPA).

3. Normal left aortic arch — plain radiographic findings
    a. The left aortic arch forms the first convexity or mogul of the left heart border on the PA radiograph.
    b. The left aortic arch produces an indentation on the left side of the distal trachea and slightly displaces the distal trachea to the right.
    c. On the lateral chest radiograph there is no retrotracheal soft tissue mass-like density in Raider's triangle above the top of the aortic arch (in other words, Raider's triangle is normal).

4. Occasionally, a normal left aortic arch has an aortic diverticulum arising from the medial aspect of the left arch distal to the origin of the LSCA.
    a. The aortic diverticulum is connected to the LPA by the ligamentum arteriosum but is not related to traction on the aortic arch by the ligamentum arteriosum.
    b. The aortic diverticulum represents persistence of the most distal part of the embryologic right aortic arch.

133

    c. Often found as an incidental finding on aortograms done for trauma

    d. Can be distinguished from a traumatic pseudoaneurysm by the fact that it has a broad base against the aortic arch and has no associated intimal flap

  5. A ductus diverticulum may also occasionally occur on the inferior aspect of the aortic arch at the location of the ligamentum arteriosum.

    a. A ductus diverticulum represents focal enlargement of a residual aortic end of the left ductus arteriosus.

    b. Ductus diverticulum is also usually an incidental finding on thoracic aortograms done for trauma.

    c. Because a ductus diverticulum can have a narrow neck, it can mimic a traumatic laceration of the aorta or a penetrating aortic ulcer.

    d. In some cases of blunt chest trauma, exploratory surgery may have to be done to distinguish a ductus diverticulum from a partial laceration of the aorta.

D. Left aortic arch with aberrant right subclavian artery (RSCA)

  1. Most common congenital anomaly of the aortic arch (0.5 –1%)

  2. The RSCA arises from the distal left arch as the fourth branch of the aorta.

    a. Many authors indicate that an aberrant RSCA is always retroesophageal in location.

    b. Others, however, indicate that this is not always the case and report the following variations on the course of an aberrant RSCA:

      1) 80% — the aberrant RSCA crosses the mediastinum behind both the trachea and esophagus to reach the right arm

      2) 15% — the aberrant RSCA crosses the mediastinum between the trachea and esophagus.

      3) 5% — the aberrant RSCA crosses the mediastinum in front of the trachea.

  3. If the aberrant RSCA arises from an aortic diverticulum, the diverticulum is called a diverticulum of Kommerell.

    a. Note, Kommerell originally described an aortic diverticulum giving rise to an aberrant RSCA in patients with a left aortic arch. Since then, the name diverticulum of Kommerell has also been applied to an aortic diverticulum giving rise to an aberrant LSCA in patients with a right aortic arch.

    b. The diverticulum of Kommerell represents persistence of a portion of the most distal part of the embryologic right aortic arch.

    c. It arises from the medial aspect of the left aortic arch distal to the LSCA and is connected to the LPA by the left ductus arteriosus and subsequently the left ligamentum arteriosum.

    d. A diverticulum of Kommerell has a smooth, broad base against the aortic arch.

    e. On the PA chest radiograph in patients with a left aortic arch, a diverticulum of Kommerell can produce a double density over the aortic knob resembling a focal aneurysm. On the lateral chest radiograph a diverticulum of Kommerell characteristically produces a retrotracheal soft tissue mass-like density in Raider's triangle. On barium swallow the "mass" is posterior to the esophagus as well.

4. Left aortic arch with aberrant RSCA is the most common cause of a vascular ring.
   a. The vascular ring formed by an aberrant RSCA is loose and seldom symptomatic.
   b. Aberrant RSCA is often discovered as an incidental finding on CT of the chest done for other reasons.
   c. Occasionally, patients with an aberrant RSCA have symptoms.
      1) Respiratory symptoms are apparently more common than esophageal symptoms and usually consist of stridor and wheezing.
      2) Dysphagia is uncommon.
      3) Dysphagia secondary to an aberrant RSCA is called dysphagia lusoria.
   d. Although aberrant RSCA infrequently causes symptoms, because it is so common, it accounts for 5-20% of vascular rings requiring surgical correction and is approximately the fourth most common cause of a vascular ring requiring surgery.
5. Radiographic findings
   a. In children the chest radiograph is often normal.
   b. In adults the chest radiograph is frequently abnormal. On the PA chest radiograph the aberrant RSCA displaces the left lung away from the top of the aortic knob. Because of its oblique orientation through the mediastinum, an aberrant RSCA causes the aortic knob to have an oblique or even flat top instead of a round top. As it exits the right side of the mediastinum, an aberrant RSCA may cause a slight convexity in the right paratracheal area.
   c. On the lateral chest radiograph an aberrant RSCA most often causes a small retrotracheal soft tissue mass-like density in Raider's triangle just above the top of the aortic arch.
   d. On barium swallow an aberrant RSCA characteristically causes an oblique extrinsic compression on the posterior aspect of the esophagus that is higher on the right than on the left.
   e. If the aberrant RSCA arises from a diverticulum of Kommerell the PA view often shows a mass-like density several centimeters in diameter along the superomedial aspect of the aortic knob, and the lateral chest radiograph shows a retrotracheal soft tissue mass-like density in Raider's triangle that is several centimeters in diameter. The diverticulum of Kommerell usually displaces the trachea anteriorly.
   f. The origin of the RSCA as the fourth branch of the left aortic arch, and presence of a diverticulum of Kommerell, can be easily established by either CT or MR. Angiography is seldom necessary to make the diagnosis.
   g. On CT and MR the diverticulum of Kommerell is seen as a broad-based focal enlargement of the medial aspect of the posterior portion

of the aortic arch that is both retrotracheal and retroesophageal in location.

6. In coarctation of aorta
   a. Aberrant RSCA may arise distal to the coarcted segment.
   b. When this happens blood flows from the right arm to the descending aorta and the usual collaterals do not develop on the right side.
   c. Patients develop unilateral inferior rib notching on the left side only.

E. Left aortic arch with aberrant right innominate artery

1. The innominate artery arises distally from the aortic arch and passes through the mediastinum anterior to the trachea to reach the right side.
2. The aberrant innominate artery forms a vascular ring that compresses the anterior aspect of the trachea.
   a. Almost always associated with severe respiratory symptoms
      1) Stridor — 100%
      2) Wheezing — 100%
      3) Cyanosis — 69%
      4) Apnea — 61%
      5) Recurrent pneumonia — 31%
   b. Although left aortic arch with an aberrant right innominate artery is an uncommon cause of a vascular ring, it is the third most common cause of a symptomatic vascular ring requiring surgery, accounting for about 8-25% of surgically treated cases.
   c. Can be associated with tracheomalacia that can cause continued respiratory symptoms postoperatively
3. Does not compress the esophagus and is not associated with any esophageal symptoms
4. Treatment consists of suturing the innominate artery to the anterior chest wall.

F. Left aortic arch with right ductus arteriosus

1. Very rare
2. Differs from a normal left aortic arch in that the right ductus arteriosus persists rather than the left
3. The right ductus arteriosus, and subsequently the right ligamentum arteriosum, connects the right pulmonary artery (RPA) to either the innominate artery or the proximal descending aorta.
4. In those cases in which the right ductus connects the RPA to the proximal descending aorta, a vascular ring is formed. This vascular ring can be symptomatic.

G. Left aortic arch with right descending aorta

1. Very rare
2. The aortic arch passes to the left of the trachea and esophagus, then passes through the mediastinum posterior to the trachea and esophagus to descend on the right side of the spine.
3. The aortic arch compresses the posterior aspects of the trachea and esophagus as it passes through the mediastinum and produces a mass-like density in Raider's triangle.

    4. Forms a vascular ring that can be symptomatic, particularly when there is also a persistent right ligamentum arteriosum.

    5. May also be associated with:

       a. Aberrant RSCA

       b. Tetralogy of Fallot (TOF)

       c. Aortic stenosis

       d. Extralobar pulmonary sequestration

H. Right aortic arch — 5 types

    1. Type 1 right aortic arch (mirror-image branching common type)

    2. Type 2 right aortic arch (mirror-image branching rare type)

    3. Type 3 right aortic arch (aberrant LSCA type)

    4. Type 4 right aortic arch (aberrant left innominate artery type)

    5. Type 5 right aortic arch (isolation of the LSCA)

    6. Types 2, 4, and 5 are very rare.

I. Type 1 right aortic arch (mirror-image branching common type)

    1. Interruption occurs in the embryologic left aortic arch distal to the left ductus arteriosus. The anterior remnant of the left aortic arch becomes the left innominate artery which gives rise to the LCCA and LSCA.

    2. The aortic arch passes to the right of the trachea and esophagus.

    3. The descending thoracic aorta is positioned to the right of the thoracic spine.

    4. Mirror-image right aortic arch has three branches — a left-sided innominate artery, RCCA, and RSCA in that order. This is an exact inversion or mirror-image of normal.

    5. A left ligamentum arteriosum connects the LPA to the LSCA anterior to the trachea.

    6. Since there is no aortic diverticulum, and the left ligamentum arteriosum connects the LPA to the LSCA anterior to the trachea rather than coursing posteriorly around the trachea and esophagus to connect it to the proximal descending aorta, the common type mirror-image right aortic arch does not form a vascular ring.

    7. Radiographic findings

       a. On the PA view there is a convexity in the right paratracheal area due to the right aortic arch. The right aortic arch produces an indentation on the right side of the distal trachea and slightly displaces the distal trachea to the left.

       b. There is no diverticulum of Kommerell and the course of the LSCA is not aberrant. On the lateral chest radiograph there is no retrotracheal soft tissue mass-like density in Raider's triangle above the top of the aortic arch (in other words, Raider's triangle is normal). This is a crucial finding in patients with right aortic arch because those with mirror-image branching (right aortic arch with no retrotracheal soft tissue density in Raider's triangle) almost always have associated cyanotic heart disease, while those with an aberrant LSCA or diverticulum of Kommerell giving rise to an aberrant LSCA (right aortic arch with retrotracheal soft tissue density in Raider's triangle) seldom have associated cyanotic heart disease.

    c. Barium swallow shows no evidence of posterior indentation on the esophagus.

    d. Angiography, CT, or MR shows a right aortic arch with mirror-image branching with no diverticulum of Kommerell. In many cases the diagnosis can be made from plain films alone. MR is currently the preferred imaging modality for confirming the diagnosis.

  8. Almost always associated with cyanotic heart disease

    a. Truncus arteriosus — mirror-image right aortic arch in 40%

    b. TOF — mirror-image right aortic arch in 25-30%

    c. Tricuspid atresia — mirror-image right aortic arch in 5-10%

    d. Transposition of the great vessels — mirror-image right aortic arch in 3%

    e. Because TOF is much more common than truncus arteriosus, most of the mirror-image right aortic arches you will encounter will be in patients with TOF.

    f. Rarely mirror-image right aortic arch is associated with ventricular septal defect (VSD).

J. Type 2 right aortic arch (mirror-image branching rare type)

  1. Extremely rare

  2. There is interruption of the embryologic left aortic arch between the left ductus arteriosus and the LSCA. The anterior remnant of the left aortic arch forms a left innominate artery resulting in a right aortic arch with mirror-image branching. A small remnant of the distal left aortic arch persists and forms an aortic diverticulum that arises from the medial aspect of the right aortic arch distal to the RSCA. A left ductus arteriosus, and subsequently a left ligamentum arteriosum, connects the LPA to the aortic diverticulum posterior to the trachea and esophagus, forming a vascular ring that can be symptomatic.

  3. The descending thoracic aorta is usually right-sided.

  4. The PA chest radiograph shows a right aortic arch. On the lateral view there is a large retrotracheal soft tissue density in Raider's triangle from the aortic diverticulum. Barium swallow shows that this soft tissue density is also retroesophageal. Angiography, CT, or MR shows a right aortic arch with mirror-image branching and an aortic diverticulum.

  5. Unlike the common type of right aortic arch with mirror-image branching, which has a high association with cyanotic heart disease, the rare type with an aortic diverticulum and associated vascular ring is usually not associated with cyanotic heart disease.

K. Type 3 right aortic arch (aberrant LSCA type)

  1. Most common type of right aortic arch (0.1% of the population).

  2. Also known as Neuhauser's anomaly

  3. There is interruption of the embryologic left aortic arch between the LSCA and the LCCA. The LCCA becomes incorporated into the anterior portion

of the right aortic arch, while the LSCA becomes incorporated into the distal portion of the right aortic arch.

4. Four branches arise from the right aortic arch — LCCA, RCCA, RSCA, and aberrant LSCA in that order.

5. The aberrant LSCA arises either directly from the posterior aspect of the right arch or from a diverticulum of Kommerell.

6. The left ductus arteriosus, and subsequently the left ligamentum arteriosum, connects the LPA either to the aortic diverticulum or the aberrant LSCA posterior to the trachea and esophagus, completing a vascular ring.

   a. The vascular ring formed by the aberrant LSCA type right aortic arch is loose and usually asymptomatic.

   b. However, some patients have symptoms including stridor, wheezing, and occasionally dysphagia.

   c. Right aortic arch with aberrant LSCA is the second most common cause of a symptomatic vascular ring requiring surgical treatment, accounting for 8-33% of surgically treated cases.

7. The descending thoracic aorta can descend on either the right or left side.

8. Radiographic findings

   a. The diagnosis can usually be made from plain films alone. On the PA view there is a convexity in the right paratracheal area due to the right aortic arch. The right aortic arch produces an indentation on the right side of the distal trachea and slightly displaces the distal trachea to the left.

   b. If there is a diverticulum of Kommerell giving rise to the aberrant LSCA, the PA chest radiograph may also show the diverticulum as a mass-like density arising from the medial aspect of the right aortic arch that may project to the left of the trachea. On the lateral chest radiograph the aortic diverticulum produces a prominent retrotracheal soft tissue mass-like density in Raider's triangle above the top of the aortic arch. The diverticulum of Kommerell usually also displaces the trachea anteriorly.

   c. Barium swallow shows a large posterior indentation on the esophagus from the aortic diverticulum.

   d. Angiography, CT, or MR shows a right aortic arch with an aortic diverticulum that gives rise to an aberrant LSCA as the fourth branch of the right aortic arch. This diverticulum, the diverticulum of Kommerell, is seen as a broad-based focal enlargement of the medial aspect of the posterior portion of the right aortic arch that is both retrotracheal and retroesophageal in location. MR is currently the preferred imaging modality for confirming the diagnosis.

   e. If there is no aortic diverticulum, the PA chest radiograph may show only the right aortic arch; however, the lateral chest radiograph still usually demonstrates a retrotracheal mass-like density due to the aberrant LSCA itself. The size of the retrotracheal density produced by an aberrant LSCA alone is usually smaller than that produced by a diverticulum of Kommerell. Barium swallow shows a posterior impression from the aberrant LSCA coursing behind the esophagus. Angiography, CT, or MR shows a right aortic arch with an aberrant LSCA arising as the fourth branch of the right aortic arch.

9. Only 5% of patients with aberrant LSCA type right aortic arch have associated cardiac defects.
    a. In the majority of cases the associated cardiac defects are of the acyanotic variety.
    b. Rarely, aberrant LSCA type right aortic arch occurs in TOF.

L. **Type 4 right aortic arch (aberrant left innominate artery)**
    1. Very rare
    2. There is interruption of embryologic left aortic arch between the ascending aorta and the LCCA. The posterior remnant of the left aortic arch becomes the left innominate artery and gives rise to the LCCA and LSCA.
    3. The RCCA and RSCA are the first two branches of the right aortic arch. The left innominate artery arises from the right aortic arch as its third branch and passes posterior to the trachea and esophagus. A left ductus arteriosus, and subsequently a left ligamentum arteriosum, connects the LPA to the aberrant left innominate artery, completing a vascular ring that may be symptomatic.
    4. The descending thoracic aorta is usually right-sided.
    5. PA and lateral chest radiographs show a right aortic arch with a retrotracheal soft tissue mass-like density in Raider's triangle. Barium swallow shows a retroesophageal defect as well. Angiography, CT, or MR show a right aortic arch with an aberrant left innominate artery passing behind the trachea and esophagus.

M. **Type 5 right aortic arch (isolation of the LSCA)**
    1. Extremely rare
    2. May be associated with TOF and other cyanotic heart defects, but in many cases isolation of the LSCA is an isolated anomaly.
    3. There is interruption of embryologic left aortic arch at two locations.
        a. Between the LCCA and LSCA
        b. Distal to the LSCA
    4. The LCCA, RCCA, and RSCA arise independently from the aorta.
    5. The LSCA is not connected to the aorta but arises from the LPA via the ductus arteriosus. When the ductus arteriosus closes the only source of blood supply to the left arm is via retrograde flow through the left vertebral artery. As a result, many patients with isolation of the LSCA eventually develop clinical signs and symptoms of subclavian steal syndrome.
    6. The right arch is anterior to the trachea and no vascular ring is formed.
    7. The descending thoracic aorta is usually right-sided.
    8. The diagnosis is usually made by angiography performed to evaluate symptoms of subclavian steal syndrome in a patient with a right aortic arch.

N. Double aortic arch
  1. Type 1 — both arches patent and functioning
     a. Most common form of double aortic arch accounting for 60-90% of cases
     b. Represents persistence of the embryologic right and left aortic arches
        1) The right aortic arch is usually larger and slightly higher in position than the left aortic arch (dominant right arch) in 50-64% of cases.
           a) The right aortic arch passes posterior to the esophagus.
           b) The left aortic arch passes anterior to the trachea.
        2) The right and left aortic arches are equal in 25%
        3) The left arch is dominant in 20%
     c. Each aortic arch gives rise to a common carotid and subclavian artery. It is not possible for an innominate artery to be present.
     d. Type 1 double aortic arch completely encircles the trachea and esophagus forming a vascular ring.
        1) Occasionally may be asymptomatic
        2) In most cases the double aortic arch produces a very tight vascular ring with compression of both the trachea and esophagus.
           a) Stridor — 100%
           b) Wheezing — 80%
           c) Cyanosis — 60%
           d) Dysphagia — 75%
        3) Most common cause of a symptomatic vascular ring requiring surgical correction, accounting for 28-43% of all cases of surgically treated vascular rings.
     e. The descending thoracic aorta is usually on the left side.
     f. Associated cardiac defects are rare.
     g. Radiographic findings
        1) The PA chest radiograph may show an indentation on both the right and left side of the distal trachea just above the carinal level.
        2) There may be a convexity in both the right and left paratracheal regions from the double arch.
        3) The lateral chest radiograph shows a prominent retrotracheal soft tissue density in Raider's triangle indenting the posterior aspect of the trachea. This retrotracheal density is caused by the posterior aspects of the right and left aortic arches as they come together to form the descending aorta.
        4) Barium swallow demonstrates extrinsic compression of both the right and left side of the esophagus at the level of the aortic arches on the anteroposterior (AP) view and a prominent posterior impression on the esophagus on the lateral view.
        5) The diagnosis can be easily confirmed by angiography, CT, or MR. Angiography is usually not necessary. MR is currently the imaging modality of choice for confirming the diagnosis.

h. Treatment — the smaller, nondominant arch, usually the left arch, is surgically divided thus releasing the trachea and esophagus from compression by the vascular ring

2. Type 2 — right arch patent with partial atresia of the left arch
    a. The atretic portion of the left arch forms a fibrous cord without a lumen that contributes to the development of a vascular ring.
    b. There are four subtypes. All are very rare.
        1) Subtype 1
            a) The right arch is patent and functioning.
            b) Atresia lies in the distal left arch between the left ductus arteriosus and the descending aorta. The anterior portion of the left aortic arch forms a left innominate artery which gives rise to a LCCA and a LSCA.
            c) This results in a right aortic arch with mirror-image branching. A small remnant of the distal left arch persists as an aortic diverticulum projecting off the medial aspect of the distal right aortic arch.
            d) The atretic portion of the left arch forms a fibrous cord without a lumen. This fibrous cord extends from the aortic diverticulum posteriorly around the left side of the trachea and esophagus to connect anteriorly to the LSCA thus forming a complete vascular ring that is often symptomatic.
            e) The ductus arteriosus, and subsequently the ligamentum arteriosum, connects the LPA to the LSCA.
            f) The descending thoracic aorta is usually left-sided.
            g) Radiographic findings
                i)   PA and lateral chest radiographs show a right aortic arch with a retrotracheal mass effect in Raider's triangle on the lateral view from the aortic diverticulum.
                ii) Barium swallow shows bilateral extrinsic defects on the esophagus at the level of the aortic arch on the AP view — the right-sided defect from the functioning right aortic arch and the left-sided defect from the fibrous cord formed by the atretic left arch. On the lateral view there is an extrinsic retroesophageal defect from the aortic diverticulum.
                iii) Angiography, CT, or MR would show a right aortic arch with mirror-image branching and an aortic diverticulum.
        2) Subtype 2
            a) Closely resembles subtype 1 in all respects except the course of the ductus arteriosus.
            b) In subtype 2 the ductus arteriosus, and subsequently the ligamentum arteriosum, connects the LPA to the aortic diverticulum rather than the LSCA as in subtype 1.
        3) Subtype 3
            a) The right arch is patent and functioning.

    b) The left aortic arch is atretic between the LCCA and LSCA. As a result, the LCCA becomes incorporated into the anterior right arch and the LSCA arises from the distal aortic arch, usually from an aortic diverticulum.

    c) This results in a right aortic arch with an aberrant LSCA arising from an aortic diverticulum.

    d) The atretic portion of the left arch forms a fibrous cord without a lumen. This fibrous cord extends from the aortic diverticulum to the LCCA. The ductus arteriosus, and subsequently the ligamentum arteriosum, also connects the LPA to the aortic diverticulum.

    e) The atretic left arch produces a symptomatic vascular ring that can be symptomatic.

    f) The descending thoracic aorta is usually left-sided.

    g) Radiographic findings

        i) PA and lateral chest radiographs show a right aortic arch with a retrotracheal mass effect in Raider's triangle on the lateral view from the aortic diverticulum.

        ii) Barium swallow shows bilateral extrinsic defects on the esophagus at the level of the aortic arch on the AP view — the right-sided defect from the functioning right aortic arch and the left-sided defect from the fibrous cord formed by the atretic left arch. On the lateral view there is an extrinsic retroesophageal defect from the aortic diverticulum.

        iii) Angiography, CT, or MR would show a right aortic arch with an aberrant LSCA arising from an aortic diverticulum.

  4) Subtype 4

    a) Hypothetical only, no described cases in literature

    b) Atresia would involve the anterior portion of the left arch proximal to the origin of the left common carotid artery resulting in a right aortic arch with an aberrant left innominate artery.

c. Although double aortic arch with left arch atresia is much less common than type 1 double aortic arch, it is the fifth most common cause of a vascular ring requiring surgery, accounting for approximately 10% of all cases of surgically treated vascular rings.

d. Differentiation between right aortic arch and double aortic arch with left arch atresia

  1) The three subtypes of double aortic arch with left arch atresia closely resemble several types of right aortic arch.

    a) Usually, no distinction between right aortic arch and double aortic arch with left arch atresia can be made by plain films, angiography, CT, or MR. CT can occasionally demonstrate the atretic left arch as a curvilinear soft tissue structure passing anterior to the trachea and connecting the LSCA to the aortic diverticulum (type 2 double aortic arch, subtypes 1 and 2) or the LCCA to the aortic diverticulum (type 2 double aortic arch, subtype 3).

b) AP views from barium studies of the esophagus in patients with double aortic arch with left arch atresia usually show both a right lateral extrinsic compression on the esophagus from the functioning right aortic arch and a left lateral extrinsic compression on the esophagus from the fibrous cord formed by the atretic left arch. As a result, the AP views of the esophagus resemble those seen in patients with functioning double aortic arch. In patients with only a right aortic arch there is a right lateral extrinsic compression of the esophagus from the right aortic arch; however, there is no left lateral extrinsic compression at the same level.

c) Because of the fibrous cord formed by the atretic left aortic arch, patients with double aortic arch with left arch atresia are more symptomatic than those with only a right aortic arch. Therefore, symptoms of tracheal or esophageal compression favors a double aortic arch with left arch atresia.

2) In most cases the distinction between the two is made during repair of a symptomatic vascular ring.

a) In double aortic arch with left arch atresia the atretic left arch forms a fibrous cord that contributes to the development of a vascular ring. This fibrous cord is located and transected during surgical repair of the vascular ring.

b) In right aortic arch the left arch completely involutes and no residual fibrous cord is found.

O. Summary — right aortic arch and double aortic arch

1. All subtypes of double aortic arch with left arch atresia, and types 2, 4 and 5 right aortic arch are very rare.

2. Most of the cases you will encounter in practice or on oral boards will fall into one of the following 3 categories:

a. Mirror-image branching right aortic arch — common type

1) High association with cyanotic congenital heart disease

2) No vascular ring is formed.

3) PA chest radiograph shows a right aortic arch.

4) Lateral chest radiograph shows a normal Raider's triangle with no retrotracheal, retroesophageal soft tissue density.

5) Barium swallow shows no posterior impression on the esophagus.

b. Aberrant LSCA type right aortic arch

1) Very low incidence of cyanotic defects — approximately 5% of patients have acyanotic cardiac defects

2) Loose vascular ring that is only occasionally symptomatic.

3) PA chest radiograph shows a right aortic arch.

4) Lateral chest radiograph shows a retrotracheal, retroesophageal soft tissue density in Raider's triangle that displaces the trachea

anteriorly. This is usually due to a diverticulum of Kommerell but can be due to the aberrant LSCA itself.

    5) Barium swallow shows a prominent posterior impression on the esophagus.

  c. Double aortic arch with both arches functioning (type 1)

    1) Usually no associated cardiac abnormality

    2) Typically a tight vascular ring with both tracheal and esophageal symptoms — most common cause of a symptomatic vascular ring requiring surgical correction

    3) PA chest radiograph shows right and left aortic arches causing an impression on both the right and left sides of the trachea.

    4) Lateral chest radiograph shows a retrotracheal, retroesophageal soft tissue density in Raider's triangle that displaces the trachea anteriorly. This is due to the posterior aspects of the right and left aortic arches as they come together to form the descending aorta.

    5) Barium swallow shows right and left lateral impressions on the esophagus on the frontal view and a prominent posterior impression on the esophagus on the lateral view.

3. Keep in the back of your mind the fact that a patient with a right aortic arch who has symptoms of subclavian steal syndrome may have isolation of the LSCA.

P. Cervical aortic arch

1. The ascending aorta arises normally from the left ventricle (LV) and extends in such a fashion that the aortic arch is situated at the thoracic inlet or lower neck.

2. More common for the cervical aortic arch to be on the right than the left

  a. Right cervical aortic arch

    1) The ascending aorta extends superiorly from the heart into the lower right neck. The aortic arch is usually not visible on the PA chest radiograph.

    2) There is usually a prominent posterior impression on the trachea in Raider's triangle on the lateral view formed by the descending aorta as it returns from the right neck and crosses the mediastinum posterior to the trachea and esophagus to then descend along the left side of the spine.

    3) Barium swallow shows a prominent posterior and right lateral extrinsic impression on the esophagus produced by the anomalously located right aortic arch.

  b. Left cervical aortic arch

    1) On the PA view the ascending aorta is higher than normal and may produce a broad convexity in the right paratracheal area. The ascending aorta crosses the mediastinum in front of the trachea to enter the left lower neck. There is a prominent soft tissue opacity at the left lung apex formed by the aortic arch as it enters the neck.

    2) The lateral view may show that the ascending aorta posteriorly displaces the trachea as it crosses the mediastinum in front of the

trachea. Since the left cervical aortic arch courses through the mediastinum anterior to the trachea there is no extrinsic compression of the esophagus.

3. Patients often present with a pulsatile neck mass.
4. May have symptoms of a vascular ring
5. Usually no associated cardiac anomalies

Q. **Summary — clinical manifestations of vascular rings**
1. In 73% of cases clinical manifestations are purely respiratory and include stridor, wheezing, dyspnea on exertion, cyanosis, intercostal and suprasternal retractions, and occasionally recurrent pulmonary infections.
2. In 20% of cases patients have both respiratory symptoms and esophageal symptoms including dysphagia and feeding difficulty.
3. In only 6-7% of cases are the symptoms purely esophageal.
4. Because the vascular ring has compressed the developing trachea, patients may have associated tracheomalacia with some continued respiratory symptoms following surgery. This usually slowly resolves.

R. **Interruption of the aortic arch**
1. There is complete anatomic interruption of the aortic arch with no connection between the proximal and descending aorta.
2. A large patent ductus arteriosus (PDA) is usually present and supplies blood to the descending aorta. If the PDA closes, blood supply to the lower body ceases.
3. One of the causes of hypoplastic left heart syndrome
4. Presents with marked cardiomegaly and congestive heart failure early in life
5. High association with intracardiac defects
   a. All cases have a large PDA
   b. VSD
      1) Almost always present
      2) The VSD may be subpulmonic in location. The pulmonary artery and pulmonic valve override the ventricular septum.
      3) May be associated with prolapse of an aortic cusp
   c. Atrial septal defect (ASD) not uncommon
   d. Subaortic stenosis
   e. Complete transposition of the great vessels (TGV)
   f. Truncus arteriosus
   g. Aorticopulmonary (AP) window
   h. Becu complex
      1) Combination of interruption of the aortic arch, subpulmonic VSD with pulmonary artery overriding the ventricular septum, and subaortic stenosis.
      2) Should be distinguished from Taussig-Bing malformation, which is a form of double-outlet right ventricle in which the pulmonary artery can also override the ventricular septum.
         a) In Becu complex the aorta arises from the LV and the great vessels are normally related.

146

b) In Taussig-Bing malformation the aorta arises from the right ventricle (RV) and is anterior to the main pulmonary artery (MPA).

6. Also associated with DiGeorge syndrome (third and fourth pharyngeal pouch syndrome)
   a. Underdevelopment of thymus and parathyroid glands
      1) Deficient cellular immunity (T-cell deficiency)
      2) Hypocalcemia
   b. Abnormalities of the conotruncus
      1) Truncus arteriosus
      2) TOF
   c. Abnormalities of the aortic arch
      1) Right aortic arch
      2) Interruption of the aortic arch
   d. Initial symptoms often related to the cardiac defects
7. Only 5% have no associated cardiac abnormality
8. Branching pattern
   a. Type A — interruption distal to the LSCA
      1) In 43% of cases the RSCA, RCCA, LCCA, and LSCA arise from the ascending aorta as four separate branches. The descending aorta connects to the MPA via a markedly enlarged PDA.
      2) Collateral pathways are identical to coarctation of the aorta, particularly if the ductus is partially or completely closed.
      3) Older patients may develop bilateral inferior rib notching.
   b. Type B — interruption between the LCCA and LSCA
      1) In 52% a right innominate artery and LCCA arise from the ascending aorta while the LSCA arises from the descending aorta. The descending aorta connects to the MPA via a markedly enlarged PDA.
      2) Intercostal collaterals develop only on the right. On the left the LSCA is the major pathway of collateral blood flow to the descending aorta. There may be retrograde blood flow in the left vertebral artery and patients may develop subclavian steal syndrome on the left side.
      3) In older patients unilateral inferior rib notching may occur on the right side.
   c. Type C — interruption between the innominate artery and LCCA
      1) In 5% the RSCA and RCCA arise from the ascending aorta and the LCCA and LSCA arise from the descending aorta. The descending aorta connects to the MPA via a markedly enlarged PDA.
      2) Again, intercostal arterial collaterals and inferior rib notching occur only on the right.
   d. Variations of all three of these types occur when the RSCA arises aberrantly from the proximal segment of the descending aorta

(aberrant RSCA) or from the RPA by way of a right ductus arteriosus (isolation of the RSCA).

    1) Those with an aberrant RSCA do not develop rib notching on the right.

    2) Both those with an aberrant RSCA and isolation of the RSCA can develop subclavian steal syndrome on the right.

  e. In those cases with an aberrant RSCA, the aberrant RSCA often produces extrinsic compression of the posterior aspects of both the trachea and esophagus.

S. Atresia of the aortic arch

1. Many authors do not make a distinction between interruption and atresia of the aortic arch and lump them together under the heading of interruption of the aortic arch.
2. Is similar in all respects to interruption of the aorta except that a fibrous cord without a lumen connects the ascending and descending aorta

T. Aortic valve atresia

1. Also part of the hypoplastic left heart syndrome
2. There is complete closure of the aortic valve associated with marked hypoplasia of the left side of the heart.
3. Patients may appear normal at birth, but they quickly develop severe congestive heart failure during the first and second day of life. This may be accompanied by mild cyanosis. Aortic valve atresia is one of the more common causes of death from congenital heart disease in the first week of life.
4. The ascending aorta is patent but is usually markedly hypoplastic. The LV is small and hypoplastic. The RV supplies blood to the entire pulmonary and systemic circulations. There is an ASD that allows blood to flow from the left atrium (LA) to the right atrium (RA). There is a large PDA that supplies blood to the descending aorta from the MPA. The coronary arteries receive their blood supply by retrograde flow from the descending aorta through the hypoplastic ascending aorta.
5. Radiographic manifestations usually consist of cardiomegaly, enlargement of the MPA, and signs of pulmonary venous hypertension including interstitial or alveolar pulmonary edema, although overcirculation of the pulmonary vascularity may occasionally be present.

U. Coarctation of the aorta

1. Localized deformity of aortic media resulting in an infolding of the posterolateral aortic wall which causes an eccentric narrowing of the aortic lumen
  a. The aortic lumen is typically narrowed more than 50%.
  b. The gradient across the narrowed segment is greater than 10 mm Hg.
2. Twice as common in males as in females
3. Accounts for approximately 5% of cases of congenital heart disease
4. Two main forms
  a. Localized (postductal or "adult" type)
    1) Most common form

2) Coarctation occurs distal to the LSCA and just distal to the ductus arteriosus.

3) During fetal development blood flow from the ductus is directed into the aortic arch proximal to the coarcted segment. This along with blood flow from the ascending aorta prevents the development of hypoplasia of the aortic arch.

4) Patients are often asymptomatic. Coarctation may be discovered incidentally during childhood, adolescence, or later life because of hypertension, a murmur, or an abnormal chest radiograph.

b. Tubular hypoplasia (diffuse, preductal, or "infantile" type)

1) Coarctation occurs distal to the LSCA but proximal to the ductus arteriosus.

2) During fetal development blood flow from the ductus is directed into the descending aorta distal to the coarcted segment, thus bypassing the aortic arch. This diminution in blood flow through the aortic arch results in a long segment of hypoplasia of the aortic arch after the origin of the innominate artery. The hypoplasia is usually most severe between the LSCA and ductus.

3) Usually presents with congestive heart failure after the ductus closes; typically in the second or third week of life

c. Rib notching in coarctation of the aorta

1) The intercostal arteries serve as a major pathway for collateral blood flow to the descending aorta. The enlarged, tortuous intercostal collaterals cause inferior pressure erosion or notching of the ribs.

2) It takes at least 8 years for rib notching to develop to a degree to become visible on chest radiographs. Rib notching increases with increasing age; approximately 75% of adults with untreated coarctation will have rib notching.

3) When the branching pattern of the aortic arch is normal and the coarcted segment is distal to the LSCA, rib notching is bilateral. This is true whether the coarctation is postductal or preductal. The first two ribs are usually spared because the first and second intercostal arteries originate from the superior intercostal artery, which arises from the subclavian artery above the coarcted segment, and do not participate in the collateral flow. The tenth, eleventh, and twelfth ribs are also spared because their intercostal arteries do not communicate with the internal mammary arteries anteriorly. Therefore rib notching is limited to the third through ninth ribs. The most prominent rib notching usually involves the fourth through eighth ribs.

4) If there is an aberrant RSCA that originates distal to the coarcted segment, blood flows from the right arm to the descending aorta and the usual intercostal artery collaterals do not develop on the right side. In these patients rib notching occurs only on the left side. There may be retrograde blood flow in the right vertebral

artery to supply blood flow to the aberrant RSCA. Patients may develop subclavian steal syndrome on the right.

5) Occasionally the coarcted segment lies proximal to the LSCA. In these patients the LSCA serves as a major collateral and rib notching develops only on the right. There may be retrograde blood flow in the left vertebral artery to supply blood flow to the LSCA. Patients may experience subclavian steal syndrome on the left side.

6) Rarely, the coarcted segment lies proximal to the LSCA in a patient with an aberrant RSCA. In these cases rib notching does not occur at all. There may be retrograde blood flow in both vertebral arteries.

V. Summary — inferior rib notching related to cardiovascular disease
1. Bilateral inferior rib notching
   a. Coarctation of the aorta — includes both preductal and postductal coarctation as long as the coarcted segment is distal to the LSCA
   b. Type A interruption of the aorta
   c. Pseudotruncus
2. Unilateral inferior rib notching on the right
   a. Mediastinal fibrosis occluding the RPA
   b. Congenital interruption of the RPA
   c. Left aortic arch with coarctation of the aorta proximal to the LSCA
   d. Right aortic arch with coarctation of the aorta with an aberrant LSCA arising distal to the coarcted segment
   e. Types B and C interruption of the aorta
   f. Blalock-Taussig shunt on the right in a patient with a left aortic arch
   g. Pulmonary arteriovenous malformation
3. Unilateral inferior rib notching on the left
   a. Mediastinal fibrosis occluding the LPA
   b. Congenital interruption of the LPA in TOF
   c. Left aortic arch with coarctation of the aorta with aberrant RSCA arising distal to the coarcted segment
   d. Right aortic arch with coarctation of the aorta proximal to the RSCA
   e. Type A interruption of the aorta with aberrant RSCA
   f. Blalock-Taussig shunt on the left in a patient with a right aortic arch
   g. Pulmonary arteriovenous malformation
4. No rib notching develops in the following situations
   a. Coarctation of the aorta proximal to the LSCA in a patient with an aberrant RSCA
   b. Type B interruption of the aorta with an aberrant RSCA
   c. Type C interruption of the aorta with an aberrant RSCA

W. Pseudocoarctation of aorta
1. Felt to represent a mild form of coarctation which is not hemodynamically significant
   a. The luminal narrowing is less than 50%.

b. The pressure gradient across the narrowed area is less than 10 mm Hg.

c. There are no collateral vessels.

2. On the PA chest radiograph there is a double density over the aortic arch which resembles an aortic aneurysm or mediastinal mass. On the lateral view there is marked kinking of the posterior aortic arch from infolding of the posterolateral aspect of the aortic wall. The marked kinking can be confirmed by angiography, reformatted images from contrast-enhanced CT, or MR. The lack of collateral vessels in a markedly kinked aortic arch is characteristic of pseudocoarctation.

X. **Supravalvular aortic stenosis**

1. Accounts for 5% of cases of congenital aortic stenosis
2. Narrowing of ascending aorta just above the coronary arteries
3. May be focal or diffuse
   a. Most common form is an hourglass deformity of the ascending aorta — 65%.
   b. There may be a simple fibrous diaphragm of the supravalvular aorta with a small opening — 10%.
   c. There may be diffuse tubular narrowing (hypoplasia) of the ascending aorta beginning just above the coronary arteries — 25%.
4. Usually part of William's syndrome
   a. Mental retardation
   b. Hypercalcemia
   c. Elfin facies
   d. Supravalvular/peripheral pulmonic stenosis
   e. Supravalvular aortic stenosis

III. **Congenital variants and anomalies of the great veins**

A. **Normal venous anatomy**

1. The azygos vein ascends from the retroperitoneum in the right hemithorax along the right side of the spine. At the T-4 or T-5 level it courses anteriorly along the right side of the mediastinum, arching over the right mainstem bronchus to empty into the posterior aspect of the superior vena cava (SVC). The right superior intercostal vein (RSIV) empties into the superior aspect of the posterior turn of the azygos arch just as the azygos vein turns anteriorly away from the spine.

2. The hemiazygos vein ascends from the retroperitoneum in the left hemithorax along the left side of the spine. The hemiazygos vein crosses the midline at the T-8 or T-9 level to communicate with the azygos vein.

3. Above this level a smaller channel, the accessory hemiazygos vein, continues the hemiazygos system superiorly. At the T-3 or T-4 level, the left superior intercostal vein (LSIV) and accessory hemiazygos vein combine in a retroaortic position and are then continued as a single channel that courses anteriorly and superiorly around the aortic arch, LSCA, and LCCA to enter the posterior aspect of the left brachiocephalic vein (LBCV).

a. Most authors consider the horizontal venous channel formed by the combination of the LSIV and accessory hemiazygos to be a continuation of the LSIV and call it the arch of the LSIV.

b. Others consider it to be a continuation of the accessory hemiazygos vein and call it the arch of the accessory hemiazygos vein.

c. It is the vascular structure that sometimes produces an oval projection off the left lateral aspect of the aortic arch known as the "aortic nipple."

B. Azygos lobe

1. Most common anatomical variant of the great veins of the thorax
2. Occurs in approximately 1% of population
3. An azygos lobe is thought to occur when the developing lung grows faster than the azygos vein can migrate to its normal position in the mediastinum, resulting in the azygos vein being engulfed by the developing right upper lobe.

   a. The azygos vein is seen as an oval structure at the inferior end of the azygos fissure.

   b. The azygos fissure, unlike other fissures, is composed of four layers of pleura — two layers of visceral pleura and two layers of parietal pleura.

   c. The azygos vein courses through the right upper lobe to enter the posterior aspect of the SVC.

   d. The SVC is more laterally positioned to the right than usual, often making the upper mediastinum appear widened.

   e. The RSIV is also positioned further to the right than usual.

C. "Hemiazygos" lobe

1. Rarely, a similar phenomenon occurs on the left, with the arch of the LSIV (also known as the arch of the accessory hemiazygos vein) being engulfed by the left upper lobe.
2. Felson refers to this rare accessory lobe as the hemiazygos lobe and its fissure as the hemiazygos fissure.

D. Anomalous course of the SVC through the right lung

1. In most cases of scimitar syndrome, the scimitar shadow is produced by an anomalous pulmonary vein returning to the RA or one of its tributaries.
2. Very rarely, the scimitar shadow can be produced by the SVC itself as it follows an anomalous course through the substance of the right lung to enter the RA.
3. Is associated with hypogenetic lung and is part of the congenital pulmonary venolobar syndrome

E. Persistent left SVC

1. Occurs in about 0.5% of the population
2. Represents persistence of the left anterior cardinal system
3. The persistent left SVC drains the left jugular and left subclavian veins. It descends anterior to the left hilum and connects to the coronary sinus to drain to the RA.

4. A persistent left SVC produces widening of the left superior mediastinum and a well-defined vertical opacity lateral to the aortic arch on PA chest radiographs.
5. A persistent left SVC may be discovered as an incidental finding on CT examinations of the chest done for other reasons. Frequently a persistent left SVC is discovered on portable chest radiographs obtained to evaluate the location of an intravascular deep line.
6. A left-sided internal jugular or subclavian deep line, such as a Swan-Ganz catheter, will descend along the left side of the mediastinum past the aortic arch and then gently curve toward the RA as it passes through the coronary sinus. Once it reaches the RA, the catheter will turn sharply upward to pass through the tricuspid valve, RV, and MPA to have its tip in either the RPA or LPA.
7. In patients with persistent left SVC, the right SVC can be either present (more common situation) or absent (less common).
   a. In those with both a right and left SVC, a deep line inserted from the right side will follow a typical course through the right SVC while a deep line inserted from the left will follow the course described above.
   b. In those in which the right SVC is absent, a deep line inserted on the right will cross the upper mediastinum through the right brachiocephalic vein to enter the left SVC and will then follow the course described above. This is a mirror-image of normal.
8. Rarely, the persistent left SVC may drain to the LA or pulmonary veins
   a. Forms a right-to-left shunt that can be associated with cyanosis
   b. May be associated with coronary sinus ASD
9. Persistent left SVC is one of the anomalies frequently present in patients with visceral heterotaxy.

F. **Azygos continuation of an interrupted inferior vena cava (IVC)**
   1. Common component of polysplenia syndrome
   2. Can also occur in congenital pulmonary venolobar syndrome
   3. The hepatic segment of the IVC is absent. The infrahepatic portion of the IVC is very small.
   4. Systemic venous return from the lower body is primarily via the azygos vein, which is markedly enlarged.
      a. The PA chest radiograph usually shows marked enlargement of the azygos vein.
      b. On the lateral view the posterior aspect of the enlarged arch of the azygos vein may produce a rounded retrotracheal density that projects over lower third of the aortic arch.
      c. The hepatic veins still drain to the RA via the suprahepatic portion of the IVC. The lateral chest radiograph still demonstrates an IVC shadow in the majority of cases, although in some cases the IVC shadow is unapparent on the lateral chest radiograph.
      d. Can occasionally occur in a patient with an azygos lobe

G. Hemiazygos continuation of an interrupted IVC

1. In some cases, continuation of IVC flow is via the hemiazygos vein rather than the azygos vein. This is rare.
2. The hemiazygos vein in the abdominal region is markedly enlarged. Although the hemiazygos vein usually crosses the mediastinum at the T-8 or T-9 level to communicate with the azygos vein, in hemiazygos continuation of an interrupted IVC the dilated hemiazygos vein usually continues cranially via the accessory hemiazygos vein which connects to the left superior intercostal vein posterior to the aortic arch and then continues via the arch of the left superior intercostal vein (the aortic nipple) along the left side of the aortic arch to connect with the LBCV.
3. Thus in hemiazygos continuation of an interrupted IVC the chest radiograph shows a mass-like density adjacent to the lateral aspect of the aortic arch that represents a massively dilated aortic nipple. Contrast-enhanced CT can easily make the diagnosis.

References and suggested additional reading

1. Arey LB. *Developmental Anatomy a Textbook and Laboratory Manual of Embryology 7th ed.* Philadelphia: Saunders, 1965.
2. Bertolini A, Pelizza A, Panizzon G, Moretti R, Bava CL, Calza G, Tacchino A. Vascular rings and slings: diagnosis and surgical treatment of 49 patients. J Cardiovasc Surg 1987; 28:301-312.
3. Chun K, Colombani PM, Dudgeon DL, Haller JA Jr. Diagnosis and management of congenital vascular rings: a 22-year experience. Ann Thorac Surg 1992; 53:597-602.
4. Felson B. *Chest Roentgenology*. Philadelphia, Saunders, 1973.
5. Heron CW, Pozniak AL, Hunter GJS, Johnson NM. Case report: anomalous systemic venous drainage occurring in association with the hypogenetic lung syndrome. Clin Radiol 1988; 39:446-449.
6. Langlois J, Binet J-P, De Brux J-L, Hvass U, Planche C. Aortic arch anomalies. In: Fallis JC, Filler RM, Lemoine G eds. *Pediatric Thoracic Surgery*. New York: Elsevier, 1991:172-188.
7. Manning WJ, Pennell DJ. *Cardiovascular Magnetic Resonance*. New York: Churchill Livingstone, 2002.
8. McLoud TC. *Thoracic Radiology: The Requisites*. St. Louis: Mosby, 1998.
9. Miller SW. *Cardiac Radiology: The Requisites*. St. Louis: Mosby, 1996.
10. Pernkopf E. *Atlas of Topographical and Applied Human Anatomy 2nd ed.* Baltimore: Urban and Schwarzenberg, 1980.
11. Raider L, Landry BA, Brogdon BG. The retrotracheal triangle. RadioGraphics 1990; 10:1055-1079.
12. Roberts CS, Othersen HB Jr, Sade RM, Smith CD 3rd, Tagge EP, Crawford FA Jr. Tracheoesophageal compression from aortic arch anomalies: analysis of 30 operatively treated children. J Pediatr Surg 1994; 29:334-337.

13. Shuford WH, Sybers RG. *The Aortic Arch and its Malformations With Emphasis on the Angiographic Features*. Springfield: Charles C Thomas, 1974.
14. Spindola-Franco H, Fish BG. *Radiology of the Heart: Cardiac Imaging in Infants, Children, and Adults*. New York: Springer-Verlag, 1985.
15. Woodring JH, Olson MA. Computed tomography of the superior intercostal veins. J Comput Tomogr 1987; 11:327-334.

John H. Woodring, M.D.

Chapter 12.

Aneurysms of the thoracic aorta and pulmonary arteries

I. **Normal dimensions of the thoracic aorta in adults**
  A. **Aortic root**
    1. Mean value — 3.7 cm
    2. Upper limit of normal (2 standard deviations above the mean) — 4.0 cm
  B. **Ascending aorta**
    1. Mean value — 3.2 cm
    2. Upper limit of normal (2 standard deviations above the mean) — 3.7 cm
  C. **Descending aorta**
    1. Mean value — 2.5 cm
    2. Upper limit of normal (2 standard deviations above the mean) — 2.8 cm
    3. The descending aorta should never be larger than the ascending aorta.
  D. **Aortic aneurysm — generally accepted size criteria**
    1. Aneurysms in the ascending aorta are greater than 5 cm in diameter.
    2. Aneurysms in the descending aorta are greater than 4 cm in diameter.

II. **Aortic dissection (dissecting hematoma)**
  A. **Most common acute emergency involving the aorta**
    1. 2-3 times more common than rupture of an abdominal aortic aneurysm
  B. **Classification**
    1. DeBakey classification
      a. Type I
        1) Begins in the ascending aorta above the aortic valve and extends into the descending aorta
      b. Type II
        1) Limited to the ascending aorta
      c. Type III
        1) Begins distal to the left subclavian artery and extends into the descending aorta
        2) Rarely propagates in a retrograde fashion to involve the aortic arch and ascending aorta
    2. Stanford classification
      a. Has largely replaced the DeBakey classification
      b. Type A — 60%
        1) Involves ascending aorta
          a) Can be limited to ascending aorta
          b) Can extend to involve the descending aorta also
        2) Intimal tear usually originates in the ascending aorta within several centimeters of the aortic root.

3) Type A classification has prognostic and therapeutic implications because of associated high risk of:
   a) Rupture into the pericardial sac with pericardial tamponade
   b) Occlusion of coronary arteries or branch vessels of the aorta
      i) Myocardial infarction
      ii) Stroke
      iii) Hemiplegia
      iv) Paraplegia
   c) Disruption of the aortic valve with aortic regurgitation and congestive heart failure
      i) Circumferential dissection may widen the aortic root so that the valve leaflets cannot close properly.
      ii) If the dissection is asymmetric, one of the valve leaflets may be downwardly displaced so that the valve cannot close properly.
      iii) A valve leaflet may be disrupted by the false lumen resulting in a flail leaflet.
4) Treatment is usually surgical
   c. Type B — 40%
   1) Involves descending aorta only
   2) Treatment is usually medical and consists mainly of control of hypertension

C. Pathogenesis
1. Most often due to degeneration of the aortic media with loss of elastic tissue and muscle cells, often referred to as cystic medial necrosis of the aorta.
   a. Hypertension — present in 70-90% of cases
   b. Can be part of a generalized connective tissue disorder
      1) Marfan's syndrome — most common cause of aortic dissection in patients under 40 years of age
         a) Ocular manifestations
            i) Bilateral dislocation of the lens
         b) Skeletal manifestations
            i) Long, slender tubular bones with arachnodactyly
            ii) Dolicocephaly with high-arched palate
            iii) Scoliosis
            iv) Pectus excavatum
            v) Anterior bowing of the sternum due to massive dilatation of the ascending aorta (annuloaortic ectasia)
            vi) Hypermobility of joints
         c) Cardiovascular manifestations
            i) Annuloaortic ectasia
            ii) Type A aortic dissection
            iii) Aneurysmal dilatation of the main pulmonary artery (MPA) — some cases of "idiopathic" dilatation of the MPA probably represent unrecognized cases of Marfan's syndrome

iv) Aortic valvular regurgitation secondary to dilatation of the aortic annulus

v) Mitral valvular regurgitation secondary to excess length of the chordae tendineae

vi) Left-to-right shunt from intracardiac rupture of a sinus of Valsalva producing an aorticocameral fistula

2) Ehlers-Danlos syndrome

3) Osteogenesis imperfecta

4) Homocystinuria

5) Relapsing polychondritis

c. Bicuspid aortic valve with aortic stenosis

d. Coarctation of the aorta

e. Turner's syndrome

f. Pregnancy — 50% of dissections in women occur during pregnancy

g. Aortitis

1) Takayasu's arteritis

2) Giant cell arteritis

3) Ankylosing spondylitis

4) Rheumatoid arthritis

5) Reiter's syndrome

6) Psoriasis

7) Systemic lupus erythematosus

h. Idiopathic — occasionally cystic medial necrosis of the aorta may be an isolated finding with no known predisposing cause (Erdheim's syndrome)

2. Other causes of aortic dissection

a. Atherosclerosis

b. Trauma — rare cause of dissection

1) Blunt chest trauma

2) Catheterization

3) Aortic surgery

c. Syphilis — rare cause of dissection

d. Tuberculosis — rare cause of dissection

3. It is thought that in most cases the initial event is an intimal tear that allows blood to enter the aortic wall and separate the media creating a true and false lumen.

a. The entry point is usually in the ascending aorta several centimeters above the aortic root, or in the descending aorta between the left subclavian artery and ligamentum arteriosum.

b. These two points are relatively fixed and are thought to experience more mechanical stress during systole.

4. Another theory is that the primary event is rupture of the vasa vasorum of the aortic wall leading to a dissecting hematoma between the media and adventitia that may or may not rupture through the intima into the aortic lumen. This may explain the fact that in a small percentage of cases of aortic dissection no intimal tear can be found at autopsy.

John H. Woodring, M.D.

5. The false lumen is often larger than the true lumen and may compress the true lumen.
6. The false lumen may have blood flow or may be thrombosed.
7. When dissection involves the descending thoracic aorta, the false lumen usually dissects down the left side of the descending aorta and may continue into the abdominal aorta.
   a. The true lumen usually supplies
      1) Celiac axis and superior mesenteric artery
      2) Right renal artery
      3) Right common iliac artery
   b. The false lumen usually supplies
      1) Left renal artery
      2) Left common iliac artery
8. Only called a dissecting aneurysm if it meets the size criteria for an aneurysm. If not, the term dissecting hematoma is preferred.

D. Clinical findings
1. Acute aortic dissection — symptoms < 2 weeks duration
2. Chronic aortic dissection — symptoms > 2 weeks duration
3. Most patients have sudden onset of severe, tearing chest pain radiating to the arms, neck, or back. In some cases; however, the chest pain is mild.
4. In 15-20% of cases there are either no symptoms at all or there are symptoms other than chest pain.
   a. Syncope
   b. Stroke
   c. Congestive heart failure
   d. Abdominal pain
5. Pulse deficits occur in up to 50% of type A and 16% of type B dissections.
6. Hemodynamic shock occurs in 25% of cases.
7. Neurological deficits are present in 25% of cases.
8. Murmur of aortic regurgitation may be present.
9. If not treated, 80% of patients with aortic dissection will die within one year. Causes of death include:
   a. Rupture into the pericardium with pericardial tamponade
   b. Rupture into the mediastinum or pleural space
   c. Aortic regurgitation with congestive heart failure
   d. Complications of occlusion of the coronary arteries or branch vessels of the aorta

E. Imaging findings
1. Plain films
   a. Widening of the mediastinum
   b. Enlargement and irregularity of aortic contour — helpful sign if it is an acute change from prior chest radiographs
      1) May involve the entire thoracic aorta
      2) "Disparity" sign

      a) Enlarged ascending aorta with normal knob and descending aorta

      b) Enlarged knob and descending aorta with normal ascending aorta

  c. Displaced intimal calcification

    1) Because the arch of the aorta is traveling obliquely from right to left in the thorax, nondisplaced intimal calcification in the more anterior portion of the aortic arch may appear to be inwardly displaced relative to the more posterior portion of the aortic arch on posteroanterior (PA) chest radiographs. Also, because the superior vena cava (SVC) is lateral to the aorta on PA chest radiographs, nondisplaced intimal calcification in the ascending aorta may appear to be inwardly displaced because of the combined thickness of the SVC and wall of the ascending aorta. For these reasons, apparently displaced intimal calcification is a relatively useless sign of aortic dissection in the ascending aorta and aortic arch on plain films.

    2) Valuable sign in the descending aorta on plain films

    3) Normal aortic wall thickness 2-3 mm

    4) If intimal calcification is more than 5mm in from the outer edge of the descending aorta there is a high probability of dissection.

    5) Displaced intimal calcification in the descending aorta on plain films is seen in about 7% of cases of aortic dissection.

  d. Cardiomegaly, signs of pericardial effusion or tamponade, and/or left ventricular (LV) failure

  e. Hemothorax

2. Aortography

  a. 86-88% sensitive, 75-95% specific

  b. Direct signs of dissection

    1) Identification of an intimal flap

    2) Identification of a false and true lumen

  c. Indirect signs of dissection

    1) Compression of true lumen

    2) Aortic wall thickening

      a) If the false lumen is thrombosed it appears as a thickened aortic wall

      b) >2 cm thickness of aortic wall suggests thrombosed false lumen

    3) Aortic regurgitation

    4) Branch vessel abnormalities

    5) Abnormal catheter position

  d. Can demonstrate entry and reentry points

  e. Diagnosis of dissection can be difficult if:

    1) False lumen thrombosed

    2) True and false lumens equally opacified

    3) Intimal flap not tangential to X-ray beam

           4) Intimal tear proximal to catheter tip

    f. Ancillary findings such as intramural, periaortic, mediastinal, and pericardial hemorrhage cannot be identified.

  3. Transesophageal echocardiography (TEE)

    a. 96-100% sensitive, 85-97% specific

    b. Good quality TEE performed by a skilled operator is comparable to CT or MR in sensitivity and specificity for aortic dissection and is often the first test performed to confirm the diagnosis.

    c. TEE can accurately identify aortic dissection, entry sites in the intimal flap, involvement of the brachiocephalic vessels, and aortic root complications including aortic regurgitation and hemopericardium and pericardial tamponade.

    d. The main limitation of TEE is the inability to evaluate intraabdominal extension of aortic dissection. This can be addressed by transabdominal ultrasound.

  4. CT

    a. 82-96% sensitive, 86-100% specific

    b. Spiral or helical CT preferred

    c. Findings on precontrast images

        1) Internal displacement of intimal calcification

        2) Visible intimal flap seen as a linear structure slightly higher in attenuation than surrounding blood, probably due to decreased attenuation of blood from anemia

        3) High attenuation thrombosed false lumen, if acutely thrombosed

        4) Enlargement of a long segment of the aorta

        5) Pericardial, mediastinal, and/or pleural hemorrhage secondary to rupture

    d. Findings on postcontrast images

        1) Contrast-filled true and false lumen separated by intimal flap; false lumen typically seen in left lateral aspect of aorta.

        2) Delayed enhancement of false lumen due to slower flow in the false lumen

        3) Thrombosis of false lumen with opacification of the true lumen only

        4) Compression of true lumen by thrombosed false lumen

        5) Extension into branch vessels

        6) Ischemia or infarction of organs supplied by branch vessels arising from the false lumen

        7) Pericardial, mediastinal, and/or pleural hemorrhage secondary to rupture

    e. CT cannot show aortic regurgitation, but it accurately demonstrates all other features of aortic dissection.

    f. Because of the speed with which the examination can be performed, CT is still the preferred choice for confirming the diagnosis of aortic dissection in acutely ill patients.

  5. MR

    a. 96-100% sensitive, 98-100% specific

b. On spin echo sequences the intimal flap appears as a line of medium signal intensity separating signal void within the true and false lumen, or an interface between different signal intensities within the true and false lumens.

1) The true lumen usually has higher flow velocity and a complete signal void. The false lumen may demonstrate medium signal intensity due to slower flow.

2) Small linear structures projecting from the wall of the false lumen to the intimal flap (aortic "cobwebs") may be seen in 20% of cases and are a reliable marker of the false lumen.

3) Entry sites are identified as focal interruptions in the intimal flap.

c. Cine gradient-echo sequences show the intimal flap as a line of low-to-medium signal intensity separating higher signal intensity flowing blood in the true and false lumens.

1) The true lumen usually has high signal intensity because of rapid flow.

2) The false lumen tends to have medium signal intensity because of slower flow or thrombosis.

   a) Thrombus is seen as areas of low-to-medium signal intensity that do not change significantly during the cardiac cycle.

   b) Slow flow is characterized by low-to-medium signal intensity that changes during different portions of the cardiac cycle.

d. Differentiation of dissection with completely thrombosed false lumen from atherosclerotic aneurysm with mural thrombus may be difficult.

1) MR cannot demonstrate displaced intimal calcification.

2) Thickening of the aortic wall may be only sign of aortic dissection in up to 14% of cases.

   a) Thrombosed false lumen may only cause minimal compression of true lumen.

   b) T1-weighted images show hyperintense foci consistent with subacute intramural hemorrhage in the aortic wall in 80% of these cases. This is consistent with early intramural hemorrhage occurring before the development of intimal rupture and blood flow in a false lumen.

e. Except for displaced intimal calcification, MR shows all of the features and complications of aortic dissection.

f. Although MR is the best imaging method for confirming the diagnosis of aortic dissection, most centers still reserve MR for stable patients because of concerns about the relatively longer time required for image acquisition and greater difficulty in monitoring hemodynamically unstable patients compared to CT.

III. **Annuloaortic ectasia**

A. **Annuloaortic ectasia refers to pear-shaped aneurysmal dilatation of the aortic root and ascending aorta**

1. Involves all 3 sinuses of Valsalva

2. Sinotubular ridge no longer identifiable as a distinct notch above the aortic root

3. Usually >5 cm diameter — the ascending aorta may occasionally be massively dilated with only minimal findings on plain films

B. **Usually caused by cystic medial necrosis of the ascending aorta (CMNAA)**

1. Associated with systemic connective tissue disorders in two-thirds of cases
   a. Marfan's syndrome
   b. Ehlers-Danlos syndrome
   c. Osteogenesis imperfecta
   d. Homocystinuria
   e. Relapsing polychondritis

2. Unrelated to systemic connective tissue disorders in one-third of cases
   a. Aortitis
      1) Takayasu's arteritis
      2) Giant cell arteritis
      3) Ankylosing spondylitis
      4) Rheumatoid arthritis
      5) Reiter's syndrome
      6) Psoriasis
      7) Systemic lupus erythematosus
      8) Syphilis — rare cause of annuloaortic ectasia
   b. Bicuspid aortic valve with aortic stenosis
   c. Coarctation of the aorta
   d. Pregnancy
   e. Idiopathic (Erdheim's syndrome)

3. Patients who have annuloaortic ectasia without other manifestations of Marfan's syndrome, who have family members with Marfan's, sometimes are referred to as "*forme fruste* Marfan's."

4. Although the terms annuloaortic ectasia and CMNAA are often used synonymously, CMNAA can be present without dilatation of the aorta.

5. Approximately 10% of patients with systemic hypertension develop a murmur of aortic regurgitation. These patients often have some degree of dilatation of the aortic root and aortic annulus although pathologic changes of CMNAA are frequently absent.

C. **Clinical findings**

1. Primary complication is type A aortic dissection.

2. In patients with annuloaortic ectasia, particularly those with Marfan's syndrome, aortic dissection is often clinically silent.
   a. Only clinical manifestation may be massive aortic regurgitation.
   b. High incidence of sudden death due to rupture into the pericardium, mediastinum, or pleural space

3. Any patient with annuloaortic ectasia who has 3+ or 4+ aortic regurgitation should be considered to have aortic dissection until proven otherwise.

D. Imaging findings
   1. Massive dilatation of ascending aorta
      a. Right hilum overlay sign (the bifurcation of the right pulmonary artery into the right upper lobe and descending branches lies more than 1 cm within the lateral edge of the mediastinal silhouette).
      b. Ascending aorta may touch sternum on lateral view.
   2. Because the ascending aorta is positioned centrally within the mediastinum, there occasionally can be marked enlargement of the ascending aorta that is not apparent on PA chest radiographs. In these patients the only radiographic clue to the massive dilatation of the ascending aorta may be filling in of the retrosternal clear space on the lateral view by the enlarged ascending aorta.
   3. The LV is usually enlarged.
   4. MR is the preferred method of confirming the diagnosis. MR nicely demonstrates annuloaortic ectasia and also shows any associated aortic regurgitation or aortic dissection.

IV. Atherosclerotic aortic aneurysms
   A. Most common form of thoracic aortic aneurysm (82%)
   B. Pathogenesis
      1. The intima thickens with age and cholesterol is deposited beneath the intima.
      2. Cholesterol deposition eventually ruptures into the lumen producing ulceration.
      3. Ulcerated plaque becomes calcified in an attempt to heal the ulcer.
      4. Deeper portions of the lesion disrupt the elastic fibers of the media weakening the aortic wall and resulting in dilatation of the aorta.
      5. Laplace's law, which is explained by the following formula:

$$T = P \times r$$

where T is the tension on the vessel wall, P is the intraluminal pressure, and r is the radius of the vessel, states that the tension on the wall of a vessel is the product of the intraluminal pressure and the radius of the vessel. As the diameter of the aorta increases, wall tension increases for any given intraluminal pressure. This results in progressive dilatation and aneurysm formation. Eventually the vascular supply to the aortic wall via the vasa vasorum is compromised.
      6. Ultimately, the wall of the aneurysm is composed of acellular, avascular connective tissue.
      7. Atherosclerotic aneurysms are usually fusiform (80%) but may be saccular (20%).
      8. Laminated clot may form in the aneurysm.
      9. High likelihood of rupture if >10 cm diameter
   C. Clinical findings
      1. Disease of the elderly

2. M:F = 3:1 to 9:1
3. Atherosclerosis and atherosclerotic aneurysms are much more common in the aortic arch and descending thoracic aorta than in the ascending aorta.
4. Causes of atherosclerotic disease of the ascending aorta
   a. Type II hyperlipoproteinemia
      1) Calcification occurs in sinuses of Valsalva and aortic cusps.
      2) Rarely causes aortic stenosis
   b. Diabetes mellitus
   c. Syphilis
5. Most thoracic aortic aneurysms are clinically silent and are found as incidental findings on chest radiographs ordered for other reasons.
6. Clinical signs and symptoms, when present, are late findings and indicate impending rupture.
   a. Chest pain
   b. Dyspnea
   c. Dysphagia
   d. Left pleural effusion, right pleural effusion uncommon but does occur
7. 50% have hypertension, coronary artery disease, or cerebrovascular disease
8. Approximately 50% have associated abdominal aortic aneurysms (conversely, about 30% of patients with abdominal aortic aneurysms also have thoracic aortic aneurysms).
9. Untreated, mean survival from diagnosis is approximately 2.4 years, with a 5-year survival less than 20%.
   a. Death common from rupture (56%)
      1) Usually occurs in aneurysms >10cm diameter
      2) Recent increase in size common
   b. The remainder usually die from associated cardiovascular disease.

D. Imaging findings
1. Atherosclerotic aneurysms — size criteria
   a. Ascending aorta >5 cm in diameter
   b. Descending aorta >4 cm in diameter
2. Dilated lumen with thin aortic wall
3. Peripheral intimal calcification in the aneurysm wall
4. Crescentic or circumferential mural thrombus — may have calcification within the mural thrombus as well
5. Residual lumen circular or occasionally irregular — flattened lumen suggests dissection with clotted false lumen

V. Penetrating atherosclerotic ulcer
A. Pathogenesis
1. Complication of atherosclerosis that occurs when a cholesterol plaque ruptures into the lumen of the aorta allowing blood to dissect into the aortic media disrupting the internal elastic lamina.
2. May penetrate through the media and be contained only by the adventitia

a. Represents a false aneurysm (pseudoaneurysm) at this phase
b. Rarely, may rupture into the pleural space resulting in death
3. Most common in descending thoracic aorta, but can occur in the ascending aorta and abdominal aorta
4. May be multiple

B. Clinical findings
1. Sudden onset of chest pain resembling acute myocardial infarction or aortic dissection
2. Patients are usually elderly and usually have diffuse atherosclerosis, hypertension, coronary artery disease, cerebrovascular disease and peripheral vascular disease.

C. Imaging findings
1. Chest radiographs usually show diffuse or focal enlargement of the descending thoracic aorta with a widened mediastinum.
2. If the ulcer is leaking there may be pleural effusion.
3. CT shows a focal excavation within an area of mural thickening. The adjacent aortic wall is thickened and irregular, and contrast enhancement may demonstrate contrast filling a crescentic hematoma within the aortic wall.
4. MR may show a subacute hematoma in the aortic wall with high signal intensity on both T1- and T2-weighted images. As the methemoglobin is further degraded and absorbed, signal intensity will return to that of adjacent soft tissue.

VI. Luetic aortic aneurysms
A. Pathogenesis
1. Prior to 1950 syphilis was the most common cause of an aortic aneurysm (50-70%); now syphilis accounts for only a small percentage of cases and the incidence is still decreasing.
2. Luetic aortitis results in obliterative endarteritis of the vasa vasorum of the aortic media with loss of elastic fibers and smooth muscle ultimately resulting in aneurysmal dilatation of the involved portion of the aorta. Giant cells and microgummas may be present.
3. Occurs in 12% of untreated patients with syphilis
a. Occurs 10-25 years after initial infection
b. Venereal disease research laboratory (VDRL) test usually positive
c. Positive microhemagglutination assay - Treponema pallidum (MHA-TP) test
4. Predominantly involves the ascending aorta and aortic arch (70%). The descending aorta is occasionally involved (30%), and the sinuses of Valsalva are rarely involved (<1%). If the coronary ostia are involved angina pectoris may be a major feature of the disease. If the aortic valve leaflets are involved aortic regurgitation may be a prominent finding.
5. Saccular aneurysm (75%); fusiform aneurysm (25%); in the rare case involving the sinuses of Valsalva, it may be difficult to differentiate syphilis from CMNAA.

      6. Predisposed to develop mycotic aneurysms
- **B. Imaging findings**
  1. Dilated aorta with pencil-thin curvilinear calcification of the aortic media
  2. Usually most evident in ascending aorta
  3. Characteristic calcification may be obscured by heavier atherosclerotic calcification.

VII. Mycotic (infected) aortic aneurysms
A. Pathogenesis
1. Aneurysms arising from non-syphilitic infections of aortic wall
2. Misnomer — mycotic aneurysms are almost exclusively bacterial — they are seldom fungal
3. Development of mycotic aneurysm requires
   a. Source of infection
   b. Preexisting defect in wall of aorta
      1) Atherosclerosis
      2) Syphilis
      3) CMNAA
      4) Coarctation
      5) Trauma
         a) Blunt or penetrating trauma
         b) Aortic surgery
         c) Arterial catheterization
4. Primary mycotic aneurysm
   a. Not associated with a demonstrable source of infection
   b. Rare
5. Secondary mycotic aneurysm
   a. Intravascular source of infection with septic embolization to diseased intima or vasa vasorum
      1) Bacterial endocarditis (12%)
      2) Intravenous drug abuse
      3) Immunocompromised host
         a) Malignancy
         b) Alcoholism
         c) Steroids
         d) Chemotherapy
         e) Autoimmune disease
         f) Diabetes mellitus
   b. Direct spread from adjacent tuberculosis of the spine, lungs, or lymph nodes
6. Mycotic aortic aneurysms — etiologic agents
   a. Causes of bacterial endocarditis
      1) *Staphylococcus aureus* (53%)
         a) Most common cause of mycotic aneurysm overall
         b) Also most common cause of mycotic aneurysm distal to coarctation of aorta

      2) Non-hemolytic *streptococcus*
      3) *Pneumococcus*
    b. *Salmonella* (33-50%)
    c. *Gonococcus*
    d. *Listeria*
    e. *Escherichia coli*
    f. *Mycobacterium tuberculosis*
  7. Histology
    a. Destruction of intima and internal elastic lamina
    b. Varying degrees of destruction of the aortic media and adventitia
  8. Usually true aneurysm; however, false aneurysms are common in tuberculous mycotic aneurysms
  9. Location (by descending frequency)
    a. Most common in ascending aorta and sinuses of Valsalva (but can occur anywhere in thoracic aorta)
    b. Abdominal visceral artery
    c. Intracranial artery
    d. Upper or lower extremity artery

B. Clinical findings
  1. Insidious onset of malaise and fever
  2. Positive blood culture in 50%
  3. If untreated there is a very high incidence of uncontrolled sepsis and spontaneous rupture (75%).
  4. Overall mortality 67%

C. Imaging findings
  1. Usually saccular aneurysm
  2. Seldom calcified
  3. Rapid expansion typical
  4. Gas bubbles may be present in the periaortic soft tissues, or occasionally within the aneurysm itself, particularly in diabetics with mycotic aneurysms due to *E. coli*.

VIII. Aneurysms associated with non-syphilitic aortitis
A. Pathogenesis
  1. Usually associated with immune complexes deposited in the vessel wall.
  2. In the acute phase there is inflammation of the intima, and often inflammation of the media and adventitia as well. As healing ensues the damaged tissue is replaced by collagen. This results in thickening of the aortic wall and crinkling of the underlying intima resulting in a characteristic "tree bark" appearance of the intimal surface of the aorta that is seen in all types of aortitis. Round cell and occasionally giant cell infiltration of the aorta also occurs.
  3. Most commonly involves the ascending aorta, but the arch and descending thoracic aorta can be involved. The abdominal aorta is seldom involved.
  4. The earliest manifestation is often dilatation of the ascending aorta.

      a. This is a nonspecific sign of aortitis because other conditions, including aortic valvular stenosis, systemic arterial hypertension, atherosclerosis, and CMNAA can also cause dilatation of the ascending aorta.

      b. Frequently associated with dilatation of the aortic annulus (annuloaortic ectasia) and aortic valvular regurgitation thus closely resembling CMNAA

5. Aneurysms are usually fusiform but are occasionally saccular.
6. Can be a cause of aortic dissection
7. Diagnosis of aortitis usually requires a high index of suspicion in patients with:

      a. History of prior temporal arteritis or polymyalgia rheumatica

      b. Clinical signs and symptoms of peripheral vasculitis

      c. Inflammatory syndrome with low-grade fever, malaise, weight loss, and myalgia

      d. Elevated erythrocyte sedimentation rate

      e. Antineutrophil cytoplasmic antibody (ANCA)

B. Causes

1. Takayasu's arteritis

      a. M:F = 1:8

      b. Age range — 12-66 years

      c. More common in Orientals

      d. Is a giant cell type arteritis

         1) Acute phase — granulomatous infiltration of elastic fibers of the aortic media by multinucleated giant cells, lymphocytes, histiocytes, and plasma cells

         2) Chronic phase — progressive fibrosis of vessel wall, luminal constriction from intimal proliferation and thrombosis, and aneurysm formation from destruction of the elastic fibers of the media

      e. Morphologically similar to giant cell arteritis; however, Takayasu's is the only arteritis that causes stenosis of the thoracic aorta

      f. Prodromal phase associated with fever, malaise, and arthritis

      g. Chronic phase characterized by vascular occlusion — particularly prone to involve arteries of the head and neck

         1) Blindness

         2) Cerebral ischemia

         3) Paresthesias

         4) Raynaud's phenomenon

         5) Claudication of the arms and legs

         6) Great vessel steal syndromes

         7) Hypertension

         8) Absent or diminished pulses

      h. There are 4 major types.

         1) Type 1

            a) 8% of cases

       b) Involves aortic arch and its branches

       c) Usually presents with stenosis of the innominate artery, carotid arteries, and subclavian arteries (classic "pulseless" type).

       d) May have focal aneurysms of the ascending aorta

       e) A variant form presents with focal aneurysms of the ascending aorta and great vessels only.

    2) Type 2

       a) 11% of cases

       b) Involves descending thoracic aorta and abdominal aorta

       c) Stenosis of the descending thoracic aorta or abdominal aorta may present with diminished or absent pulses in the lower extremities ("atypical coarctation" type).

       d) Renal artery stenosis may result in systemic arterial hypertension.

    3) Type 3

       a) 55% of cases

       b) Has features of both types 1 and 2

    4) Type 4

       a) Has involvement of the pulmonary arteries

       b) Occurs in about half of cases

  i. May be associated with aortic root dilatation and aortic valvular regurgitation (annuloaortic ectasia) and aortic dissection

2. Giant cell arteritis

  a. M:F = 1:3

  b. Occurs in older persons, typically Caucasians; age range — 65-75 years

  c. Histology similar to that of Takayasu's arteritis

    1) Acute phase — granulomatous infiltration of elastic fibers of the arterial wall by multinucleated giant cells, lymphocytes, histiocytes, and plasma cells

    2) Chronic phase — progressive fibrosis of vessel wall, luminal constriction from intimal proliferation and thrombosis, and aneurysm formation from destruction of the elastic fibers of the media

  d. Prodromal phase of malaise, low-grade fever, weight loss, myalgia, and unilateral headache lasting 1-3 weeks

  e. Chronic phase

    1) Palpable, tender temporal artery

    2) Jaw claudication, claudication of arms and legs

    3) Raynaud's phenomenon

    4) Neuro-ophthalmic manifestations

       a) Visual impairment, blindness

       b) Diplopia

       c) Cerebral ischemia

       d) Paresthesias

       e) Great vessel steal syndromes

        5) Polymyalgia rheumatica
          a) 50% of cases
          b) Severe pain in shoulder and hip girdles
        6) Elevated erythrocyte sedimentation rate
    f. Can affect any artery but tends to involve the medium-sized branches of the aortic arch and branches of the external carotid arteries particularly the temporal arteries
        1) Long smooth areas of arterial stenosis with skip areas
        2) Smooth tapered occlusions with collaterals
    g. May be associated with aortic root dilatation and aortic valvular regurgitation (annuloaortic ectasia) and aortic dissection
    h. Diagnosis usually made by biopsy of the abnormal temporal artery

3. Ankylosing spondylitis
    a. 12% of cases develop aortitis with dilatation of aortic root and ascending aorta crossing the sinotubular ridge — develops about 15 years after onset of arthritic symptoms
    b. Inflammatory process and scarring involves the sinuses of Valsalva, both the free edge and bases of the aortic cusps, and extends below the aortic valve to involve the mitral valve as well.
        1) Results in annuloaortic ectasia and aortic valvular regurgitation
        2) Aortic regurgitation often progressive and severe
        3) Mitral regurgitation usually mild
    c. Fibrosis may extend into the ventricular septum resulting in arrhythmias and conduction abnormalities.

4. Other causes of aortitis that can be associated with dilatation of the ascending aorta and aortic regurgitation
    a. Rheumatoid arthritis — aortitis occurs in 10-15% of cases
    b. Reiter's syndrome — aortitis occurs in 10-15% of cases
    c. Relapsing polychondritis — aortitis occurs in 10-15% of cases
    d. Psoriasis
    e. Systemic lupus erythematosus
    f. Scleroderma
    g. Ulcerative colitis
    h. Rheumatic fever

5. Behçet's disease
    a. Syndrome characterized by
        1) Aphthous stomatitis
        2) Genital ulcerations
        3) Uveitis
        4) May also have erythema nodosum, arthritis, thrombophlebitis, neurological syndromes, epididymitis, orchitis, colitis, and vasculitis
    b. Vasculitis of Behçet's disease involves both the arterial and venous systems.
        1) The aorta and pulmonary arteries are the most frequently involved arteries.
        2) Aortic involvement occurs in about 2% of cases.

a) May have aneurysms of the ascending aorta with aortic valvular regurgitation resembling annuloaortic ectasia

b) Aneurysms can be either saccular or fusiform.

 3) Pulmonary involvement occurs in about 5% of cases.

  a) Consists of pulmonary arterial thrombosis, pulmonary infarction, and pulmonary artery aneurysms

  b) Pulmonary arterial thrombosis most often involves the interlobar branch of the right pulmonary artery.

  c) Pulmonary artery aneurysms occur in about 5% of cases with pulmonary involvement.

   i) Measure from 1-3 cm in diameter

   ii) Most common in the perihilar regions

   iii) May spontaneously rupture

  d) Can also have pulmonary hemorrhage and pulmonary infarction. Pulmonary hemorrhage can be massive and result in death.

 c. Vasculitis also involves the venous system (25%).

  1) Can involve the superficial and deep veins

  2) Thrombosis and occlusion of the SVC and/or brachiocephalic veins can result in widening of the mediastinum and SVC syndrome.

  3) Iatrogenic insertion of a venous catheter can result in venous thrombosis or propagation of an existing thrombus considerably worsening the patient's condition. Because of this, noninvasive imaging (CT/MR) is preferred to arteriography in evaluation of pulmonary involvement in Behçet's disease.

C. Imaging findings

 1. CT, MR, and arteriography may be performed for evaluation of aortitis.

 2. CT and MR are excellent in demonstrating aneurysms of the aorta and larger arteries and thickening of the aortic wall, which is a hallmark of aortitis. Both CT and MR may show thrombosed aneurysms that cannot be diagnosed by arteriography. CT and MR can also show luminal irregularity and narrowing in the aorta and larger arteries and can demonstrate narrowing or occlusion of smaller arteries in many cases. MR is excellent in documenting aortic valvular regurgitation, whereas CT is not.

 3. Arteriography can also document aortic valvular regurgitation and is better than CT or MR at showing luminal irregularity in the aorta and larger arteries. Furthermore, arteriography is excellent in showing narrowing or occlusion of small arteries. Unfortunately, arteriography is poor in showing thickening of the aortic wall.

IX. Poststenotic aortic aneurysms
A. Aortic valvular stenosis

 1. Poststenotic dilatation of the ascending aorta in patients with aortic valvular stenosis can be aneurysmal.

    2. Patients with coarctation of the aorta who also have a bicuspid aortic valve are particularly prone to develop aneurysms of the ascending aorta.

        a. Coarctation of the aorta increases pressure in the ascending aorta. This can cause dilatation of the ascending aorta and aortic root, resulting in annuloaortic ectasia.

        b. The increased pressure in the ascending aorta from the jet of blood flowing through a stenotic aortic valve, coupled with the increased pressure in the ascending aorta from coarctation of the aorta, can result in massive dilatation of the ascending aorta.

B. Coarctation of aorta

    1. Can be associated with annuloaortic ectasia

    2. Can occasionally have poststenotic dilatation of the descending aorta that is of aneurysmal degree.

X. Traumatic aneurysms of the thoracic aorta

  A. Causes

    1. Surgery

        a. Aortic incision

        b. Aortic cannulation

        c. Aortic clamping

    2. Penetrating trauma

    3. Blunt chest trauma

  B. Blunt traumatic injury of the thoracic aorta and brachiocephalic arteries

    1. Major causes

        a. Rapid deceleration

            1) Automobile accident (35 mph or more) — most common cause

            2) Airplane or helicopter crash

        b. Fall from a height (10 feet or higher)

        c. Crush injury

    2. Proposed mechanisms of injury

        a. Shearing effect from rapid deceleration at points of relative aortic fixation

            1) Aortic root

            2) Aortic isthmus at attachment of ligamentum arteriosum

            3) Aortic hiatus in diaphragm

        b. Crushing of the aorta, particularly near the isthmus, between the sternum and upper thoracic spine (thoracic "pinch" mechanism)

        c. Profound intraluminal hypertension at time of blunt injury

        d. Compressive force displacing the heart into the left hemithorax which will pull the ascending aorta away from the brachiocephalic vessels

        e. Distracting injury to the head, neck, or shoulders which can tear the brachiocephalic arteries away from the aorta

    3. In the majority of cases all three layers of the aortic wall (intima, media, and adventitia) are lacerated resulting in a false aneurysm (pseudoaneurysm). Occasionally both the intima and media are completely lacerated but the adventitia remains intact.

a. Most often there is complete circumferential laceration of the aorta.
   1) The proximal and distal ends of the aorta are retracted leaving a large pseudoaneurysm between the two. The pseudoaneurysm is contained only by surrounding mediastinal tissue, periaortic hematoma, and occasionally some residual adventitial tissue. The "wall" of the pseudoaneurysm typically has a consistency similar to wet tissue paper.
   2) The torn edges of the retracted proximal and distal aorta project into the contrast column producing what is erroneously referred to as an "intimal flap."
   3) The "intimal flap" seen in association with complete circumferential lacerations of the aorta is much thicker than the aortic intima and represents either the combined intima and media or all three layers of the torn aortic wall.
b. Occasionally the aorta is lacerated only along a portion of its circumference.
   1) This occurs most often along the inferior aspect of the aortic isthmus just distal to the left subclavian artery at or near the insertion of the ligamentum arteriosum.
   2) The superior aspect of the aortic arch is intact. There is a focal pseudoaneurysm along the inferior aspect of the aortic arch that is contained only by surrounding mediastinal tissues, periaortic hematoma, and occasionally some residual adventitial tissue. The torn edge of the aorta projects into the contrast column again forming an "intimal flap" that represents the torn aortic wall.
   3) Differentiation of traumatic pseudoaneurysm from partial aortic laceration from an aortic diverticulum or ductus diverticulum
      a) Traumatic pseudoaneurysms from partial aortic laceration usually occur along the inferior aspect of the aortic arch. They can have either a broad or narrow neck and often have an irregular contour and a demonstrable "intimal flap."
      b) An aortic diverticulum arises from the medial aspect of the posterior aortic arch. It has a smooth contour, a broad neck, and no "intimal flap."
      c) Differentiation of a small pseudoaneurysm from a ductus diverticulum can be difficult since both characteristically arise from the inferior aspect of the aortic arch at the location of the ligamentum arteriosum and both can have a narrow neck. Demonstration of an "intimal flap" or an irregular contour would indicate a traumatic pseudoaneurysm; however, in some cases of blunt chest trauma, exploratory surgery may have to be done to distinguish a ductus diverticulum from a partial laceration of the aorta.
c. In a small percentage of cases disruption of the aortic wall is associated with subintimal aortic dissection distal to the site of aortic laceration and pseudoaneurysm formation.

       1) The intima displaced by the aortic dissection can partially obstruct the lumen of the descending thoracic aorta resulting in upper extremity hypertension, diminished blood flow to the lower body, and diminished or absent arterial pulses in the lower extremities (acute coarctation syndrome).

       2) The dissection can also occlude branch vessels of the aorta resulting in:

         a) Stroke

         b) Hemiplegia

         c) Paraplegia

4. Rarely blunt traumatic injury to the aorta is limited to the intima.

    a. In the most minimal form the intima is intact and there is only a small area of subintimal hemorrhage.

       1) This form of aortic injury is usually found at autopsy in trauma victims who have died of other causes.

       2) Is usually of no clinical significance but can cause aortic dissection

    b. Rarely there is laceration of the intima alone resulting in a true intimal flap.

       1) Can spontaneously heal

       2) Can cause aortic dissection

       3) Usually treated medically

5. High incidence of rupture and death if untreated

    a. Accounts for 10-15% of all traffic fatalities in United States

    b. 80-90% of victims die at the accident scene

       1) A fairly high percentage of these patients have laceration of the aortic root (30%).

       2) Death usually results from either rupture of the pseudoaneurysm into the pericardium causing pericardial tamponade or rupture into the pleural space resulting in exsanguination.

    c. 10-20% survive long enough to come to medical attention

       1) 50% of those who survive the initial injury will die within the first 24 hours after injury

       2) 98% die within the first 3 months following the initial injury

       3) Only 2% of initial survivors will live long enough without treatment to develop a chronic traumatic pseudoaneurysm.

         a) This group represents only 0.3% (3/1000) of those who initially sustain traumatic aortic laceration.

         b) Eventually most of the patients who do live long enough to develop a chronic traumatic pseudoaneurysm will also die from rupture of the pseudoaneurysm.

    d. With early diagnosis and surgical correction survival has been reported to range from 70-86% with survival approaching 100% in some series.

6. Location of vascular injury (in those surviving long enough to reach medical attention)

    a. Aortic isthmus — 70-75%

b. Brachiocephalic arteries — 25%

c. Aortic root — 5%

d. At or just above aortic hiatus in diaphragm — 1%

e. Multiple sites — reported, may account for about 1% of cases

f. Remember, in blunt trauma victims dying at the scene of the accident, the percentage of aortic root injuries is higher (about 30%) and the percentage of injuries at the aortic isthmus is therefore relatively lower (about 50%). In reported series of initial survivors that do not include brachiocephalic arterial injuries in the statistics, the percentage of injuries at the aortic isthmus is relatively higher (90-95%).

7. Clinical signs and symptoms of possible thoracic aortic or brachiocephalic arterial injury.

a. Visible chest injury — 66%

b. Retrosternal or interscapular chest pain — 20%

c. Shock — 40%

d. Drop in hematocrit — 13%

e. Hemothorax — 5-10%

f. Pericardial tamponade — rare in those who survive to reach medical attention, usually occurs in patients with laceration of the intrapericardial portion of the ascending aorta

g. Abnormal systolic murmur in left anterior second intercostal space or interscapular area to left of spine — 17-30%

h. Palpable pulse deficits — 26-43%

   1) Pulse deficits in the upper extremities suggest possible brachiocephalic arterial injury.

   2) Acute coarctation syndrome with upper extremity hypertension, lower extremity hypotension, and palpable pulse deficits in the lower extremities may occur from occlusion of the descending thoracic aorta from either a displaced intimal flap or from compression of the proximal descending aorta by a periaortic hematoma associated with the vascular injury.

   3) Best clinical sign of possible aortic or brachiocephalic arterial injury

i. Hoarseness — rare, due to compression of the left recurrent laryngeal nerve by a pseudoaneurysm arising from the aortic isthmus

j. Dyspnea — 7%

k. Dysphagia — rare, due to esophageal compression by a pseudoaneurysm arising from the aortic isthmus

l. Unfortunately, most of these signs and symptoms also occur in victims of blunt chest trauma who do not have aortic or brachiocephalic arterial injury. Of these, only palpable pulse deficits occur more frequently in patients with aortic or brachiocephalic arterial injury than in blunt trauma victims who do not have aortic or brachiocephalic arterial injury.

8. Radiographic signs associated with traumatic laceration of the thoracic aorta and brachiocephalic arteries

a. Common signs

1) Widened upper mediastinum
   a) Considered to be the "hallmark" of traumatic laceration of the thoracic aorta and brachiocephalic arteries
   b) Transverse mediastinal width
      i) Transverse mediastinal width can be determined by assessing the midline-to-right (MR) distance and midline-to-left (ML) distance of the upper mediastinum.
      ii) Draw a vertical line over the midline of the spine (the spinous processes) at the level of the aortic arch. The MR distance is defined as the distance from the midline to the point where the SVC crosses the right upper lobe bronchus. The ML distance is the distance from the midline to the point where the left subclavian artery arises from the aortic arch.
      iii) Transverse mediastinal width (MR + ML) is less than 7.5 cm in 95% of normals.
      iv) Transverse mediastinal width is 7.5 cm or more in 59% of patients with traumatic laceration of the thoracic aorta or brachiocephalic arteries. Unfortunately, transverse mediastinal width is <7.5 cm in 41% of patients with traumatic laceration of the thoracic aorta or brachiocephalic arteries.
      v) Reducing the accepted value for a normal transverse mediastinal width to 5.5 cm detects 100% of patients with vascular injury but results in a 74% false-positive rate.
   c) Mediastinal-width to chest-width (M/C) ratio
      i) The M/C ratio is determined by dividing the maximum transverse width of the mediastinum at the level of the aortic knob (M) by the internal diameter of the thorax from inner rib to inner rib at the level of the aortic knob (C).
      ii) The M/C ratio is less than 0.38 in 95% of normals.
      iii) The M/C ratio is 0.38 or more in 31% of patients with traumatic laceration of the thoracic aorta or brachiocephalic arteries. Unfortunately, the M/C ratio is <0.38 in 69% of patients with traumatic laceration of the thoracic aorta or brachiocephalic arteries.
      iv) Reducing the accepted value for a normal M/C ratio to 0.25 detects 100% of patients with vascular injury but results in an 87% false-positive rate.
   d) While using a transverse mediastinal width of 7.5 cm or more or an M/C ratio 0.38 or more can identify many cases of traumatic injury to the thoracic aorta and brachiocephalic arteries (59% and 31%, respectively), reliance on these measurements of mediastinal width alone can result in many cases being missed. Because of this, the diagnosis of mediastinal hemorrhage and possible associated traumatic injury of the thoracic aorta or

brachiocephalic injuries is based primarily upon alterations of mediastinal anatomy including the following:

2) Indistinct or enlarged aortic knob
   a) Generally best sign of possible vascular injury
   b) Present in 75% of cases — typically those with injury of the aortic isthmus
   c) Often absent, however, when the vascular injury involves the ascending aorta, distal descending thoracic aorta at the level of the aortic hiatus in the diaphragm, or the brachiocephalic arteries

3) Obscuration of the aortopulmonary window
   a) Present in about 70% of cases — typically those with injury of the aortic isthmus
   b) Often absent, however, when the vascular injury involves the ascending aorta, distal descending thoracic aorta, or the brachiocephalic arteries

4) Widened right paratracheal stripe
   a) Present in approximately 66% of cases
   b) May be seen in traumatic laceration of the ascending aorta, aortic isthmus, innominate artery, or right subclavian artery

5) Apical cap sign
   a) Occurs in about 45% of cases
   b) Occurs in both aortic and brachiocephalic arterial injury
   c) May be the earliest sign of mediastinal hemorrhage and sometimes may be the only sign of mediastinal hemorrhage
   d) Can also be a sign of traumatic avulsion of the brachial plexus
   e) False-positives occur secondary to preexisting apical pleural thickening.

6) Deviation of the trachea to the right of midline (spinous process of T-4 is used as midline indicator)
   a) Occurs in about 40% of cases
   b) Usually due to a large pseudoaneurysm of the aortic isthmus and/or associated periaortic hematoma displacing the trachea to the right
   c) Usually absent when the vascular injury involves the ascending aorta, distal descending thoracic aorta, or the brachiocephalic arteries — in fact, in some cases of injury to the ascending aorta, innominate artery, or right subclavian artery the trachea may be displaced to the left
   d) False-positives occur secondary to atherosclerotic ectasia of the thoracic aorta and rotation of the patient to the right.

7) Deviation of nasogastric tube to the right of midline (spinous process of T-4 is used as midline indicator)
   a) Also occurs in about 40% of cases
   b) A nasogastric tube must be placed in order to see this sign.

      c) Again, usually due to a large pseudoaneurysm of the aortic isthmus and/or associated periaortic hematoma displacing the esophagus to the right

      d) May be absent when the vascular injury involves the ascending aorta, distal descending aorta, or the brachiocephalic arteries

      e) False positives occur if the tip of the nasogastric tube is directed against the greater curvature of the stomach.

  8) Downward displacement of the left mainstem bronchus

      a) Present in roughly 25% of cases

      b) Usually due to a large pseudoaneurysm of the aortic isthmus and/or associated periaortic hematoma displacing the left mainstem bronchus downward

      c) Usually absent when the vascular injury involves the ascending aorta, distal descending thoracic aorta, or the brachiocephalic arteries

  9) Widened paraspinal interfaces

      a) Occurs in roughly 25% of cases

      b) Normally there is no discernible right paraspinal interface. Therefore, a visible right paraspinal interface is suspicious for mediastinal hemorrhage in a blunt chest trauma victim.

      c) The normal left paraspinal interface is produced by the descending thoracic aorta, which elevates the left paraspinal pleural reflection away from the thoracic vertebral column. Therefore, a left paraspinal interface is normally seen from the level of the aortic knob down to the diaphragm. The left paraspinal interface, however, does not normally extend above the level of the aortic knob. The left paraspinal interface is usually smooth and less than 1 cm wide. Marked widening of the left paraspinal interface, or extension of the left paraspinal interface above the level of the aortic knob is abnormal and is suspicious for mediastinal hemorrhage in a blunt chest trauma victim.

      d) False-positives occur frequently due to fractures of the thoracic spine.

  10) Left hemothorax

      a) Is more suggestive of vascular injury than right hemothorax

      b) Suggests possible vascular injury if it is large or if it recurs after chest tube drainage

      c) Occurs in about 5-10% of cases

b. Uncommon signs

  1) Right lateral displacement of the SVC

  2) Displaced intimal calcification in the aorta

  3) Enlarged cardiac silhouette secondary to hemopericardium

  4) Anterior displacement of the trachea on the lateral view

  5) Obscuration of the azygos vein

   c. Overall, approximately 93% of the cases of traumatic laceration of the aorta and brachiocephalic arteries will be detected on the initial chest radiograph because of the presence of one or more of the mediastinal abnormalities listed above.

9. Accessory clinical and radiographic signs that may increase the yield of diagnosis of traumatic laceration of the thoracic aorta and brachiocephalic arteries in patients with a normal mediastinum on initial chest radiograph
   a. Sternal fracture
   b. Posteriorly displaced clavicular fracture
   c. Multiple rib fractures with crushed chest
   d. First or second rib fractures
   e. Brachial plexus palsy
   f. Diminished or absent pulses in the upper or lower extremities
   g. Acute coarctation syndrome
   h. Palpable supraclavicular hematoma
   i. Unexplained hypotension
   j. These accessory signs will allow detection of an additional 5-6% of the cases of acute traumatic laceration of the thoracic aorta and brachiocephalic arteries. In other words, 98-99% of all cases of traumatic laceration of the thoracic aorta and brachiocephalic arteries are detected in the emergency room because of the presence of one or more clinical or radiographic signs of vascular injury.

10. It should be noted, however, that 1-2% of cases of traumatic laceration of the thoracic aorta and brachiocephalic arteries may have no clinical or radiographic evidence of vascular injury initially.

11. Delayed presentation of traumatic laceration of the thoracic aorta and brachiocephalic arteries
   a. Of the 1-2% of cases that initially have no clinical or radiographic signs of vascular injury at all, or that have very subtle signs that do not initially arouse clinical suspicion such as an isolated small apical cap, most will develop progressive mediastinal abnormality on their chest radiographs from expansion of the pseudoaneurysm and/or contained leakage of blood into the mediastinum over the next 2 or 3 days following admission.
   b. This phenomenon is important to recognize for several reasons.
      1) Many physicians are not aware of its existence. Therefore, they may erroneously assume that if a blunt chest trauma victim has no clinical or radiographic evidence of vascular injury on initial examination, vascular injury is completely excluded.
      2) There is a high incidence of rupture of the pseudoaneurysm and death, so identifying delayed presentation of traumatic laceration of the thoracic aorta or brachiocephalic arteries may save the patient's life.
      3) It has been on oral boards (number 2 is the real reason this is important, but despite the fact that some physicians say this

phenomenon does not exist, you may get a case on oral boards anyway!).

12. Chronic traumatic aortic pseudoaneurysm

 a. A very small number of cases of traumatic laceration of the thoracic aorta escape clinical detection in the period of time immediately following the traumatic event. These patients may present with a chronic traumatic pseudoaneurysm of the aorta months, years, or even decades later.

 b. A chronic aortic pseudoaneurysm is generally defined as a pseudoaneurysm of the aorta that has been present for more than 3 months.

 c. More common with partial laceration of the aorta

 d. If the patient survives for several weeks a fibrous wall begins to form around the periphery of the pseudoaneurysm. This fibrous wall is fairly well formed after 3 months; however, it is not sufficiently strong to prevent rupture of the pseudoaneurysm. Eventually almost all patients with chronic traumatic pseudoaneurysms of the aorta will die from rupture of the pseudoaneurysm into the pleural space.

 e. Chronic traumatic pseudoaneurysms are usually discovered incidentally on chest radiographs obtained for other reasons.

 f. There may be other signs of prior blunt chest trauma including old clavicle, rib, and/or sternal fractures.

 g. Chronic traumatic pseudoaneurysms most often appear as a well-rounded "mass" adjacent to the aortic isthmus and often contain thin, peripheral, rim calcification.

 h. Need to be surgically repaired since almost all will eventually rupture resulting in death

13. Diagnostic confirmation of aortic/brachiocephalic arterial injury

 a. Aortography remains the gold-standard of diagnosis.

 b. Fast CT techniques (spiral CT or multi-slice helical CT) have largely supplanted aortography in the initial evaluation of major thoracic arterial injury, with aortography increasingly being reserved for questionable cases.

 c. Findings are similar for both aortography and CT.

  1) Complete laceration

   a) The proximal and distal ends of the aorta are retracted leaving a large pseudoaneurysm between the two.

   b) The pseudoaneurysm is usually larger in diameter than the torn aorta and has an irregular or lobular outline.

   c) The torn edges of the retracted proximal and distal aorta project into the contrast column producing "intimal flaps" at both ends of the pseudoaneurysm.

  2) Partial laceration

   a) There is a focal defect in the wall of the aorta allowing contrast in the lumen to extend through the aortic wall into a periaortic pseudoaneurysm that has an irregular or lobular outline.

         b) The pseudoaneurysm can have a narrow or broad neck.

         c) The torn edge of the aorta projects into the contrast column forming an "intimal flap."

    3) Occasionally there is aortic dissection distal to the pseudoaneurysm, and rarely there is aortic dissection only. Very rarely there is only an isolated intimal flap without dissection.

    4) In chronic traumatic pseudoaneurysms there may be mural thrombus along the wall of the pseudoaneurysm and there is often peripheral rim calcification in the wall as well. CT is better than aortography in demonstrating these features of chronic pseudoaneurysms.

XI. Congenital aortic aneurysms
A. Exceedingly rare
B. Usually involve either a ductus diverticulum or sinus of Valsalva

XII. Sinus of Valsalva aneurysms
A. Causes
1. Congenital
   a. Associated with coarctation
   b. Marfan's syndrome
   c. Ehlers-Danlos syndrome
2. Endocarditis of aortic valve
3. Mycotic
4. Luetic
5. Atherosclerotic
6. Dissection

B. Imaging findings
1. Because the sinuses of Valsalva are buried deep within the cardiac silhouette, sinus of Valsalva aneurysms usually do not show on chest radiographs.
   a. Occasionally a large aneurysm of the left sinus of Valsalva may present as an abnormal convexity in the coronary artery triangle.
      1) May mimic an enlarged left atrial appendage
      2) Can have peripheral rim calcification
   b. Rarely a large aneurysm of the right sinus of Valsalva may present as a focal bulge along the right heart border near the junction of the SVC and right atrium.
      1) Can have peripheral rim calcification
2. The diagnosis is usually made by aortography or MR.

XIII. Pulmonary artery aneurysms
A. Causes
1. Congenital
   a. Post-stenotic dilatation of MPA secondary to valvular pulmonic stenosis

        b. Post-stenotic dilatation of MPA or the right or left pulmonary arteries or their branches secondary to supravalvular/peripheral pulmonic stenosis

        c. Marfan's syndrome

        d. *Forme fruste* Marfan's

    2. Traumatic (pseudoaneurysm)

        a. Instrumentation, catheterization of pulmonary arteries

           1) Most commonly caused by perforation of a pulmonary artery by a Swan-Ganz catheter

        b. Penetrating trauma

        c. Blunt chest trauma

    3. Infectious (mycotic)

        a. Pyogenic bacteria are now the most common cause of mycotic aneurysms of the pulmonary arteries.

           1) *Staphylococcus aureus*

           2) *Streptococcus*

        b. Fungi — rare cause of pulmonary artery aneurysm

           1) *Aspergillus*

           2) *Candida*

        c. Tuberculosis (Rasmussen aneurysm) — once common, now rare

        d. Syphilis — once common, now rare

    4. Immunologic

        a. Behçet's disease

        b. Takayasu's arteritis type 4

           1) Pulmonary arterial lesions seen in Takayasu's arteritis

               a) Dilatation of MPA (20%)

               b) Nodular thrombi (3%)

               c) Narrowing of pulmonary arteries with "pruned tree" appearance (66%)

           2) May have systemic artery-to-pulmonary artery shunts

    5. Secondary to pulmonary arterial hypertension

    6. Hughes-Stovin syndrome

        a. Association of pulmonary artery aneurysms and venous thrombosis without the oral and genital ulcers, uveitis, and other manifestations of Behçet's disease

        b. Cause unknown, may be a *forme fruste* of Behçet's disease

B. Clinical findings

    1. Patients usually asymptomatic

    2. Hemoptysis suggests impending rupture but can also occur from Behçet's disease or Takayasu's arteritis type 4.

C. Imaging findings

    1. Pulmonary artery aneurysms are usually seen as sharply circumscribed, rounded masses adjacent to the MPA or hilum.

    2. Occasionally may be away from the hilum and resemble a solitary pulmonary nodule or mass

3. The diagnosis can be made by CT, MR, or pulmonary arteriography. CT and MR are preferred, particularly in patients with Behçet's disease.

References and suggested additional reading

1. Abrams HL, ed. *Angiography 2nd ed.* Boston: Little, Brown, 1971.
2. Dähnert W. *Radiology Review Manual 4th ed.* Philadelphia: Lippincott Williams & Wilkins, 2000.
3. Felson B. *Chest Roentgenology.* Philadelphia: Saunders, 1973.
4. Green CE, Elliott LP. The chest film in aortic valve regurgitation. In: Taveras JM, Ferrucci JT eds. *Radiology: Diagnosis-Imaging-Intervention.* Volume 2, Chapter 46. Hagerstown: Lippincott, 1991.
5. Guthaner DF. The plain chest film in assessing aneurysms and dissecting hematomas of the thoracic aorta. In: Taveras JM, Ferrucci JT eds. *Radiology: Diagnosis-Imaging-Intervention.* Volume 2, Chapter 33. Hagerstown: Lippincott, 1991.
6. Hachulla E, Beregi JP. Diagnosis of aortitis. J Mal Vasc 2001; 26:223-227.
7. Hartnell GG. Imaging of aortic aneurysms and dissection: CT and MRI. J Thorac Imaging 2001; 16:35-46.
8. Keats TE, Sistrom C. *Atlas of Radiologic Measurement, 7th ed.* St. Louis: Mosby, 2001.
9. Lee JKT, Sagel SS, Stanley RJ, Heiken JP. *Computed Body Tomography with MRI Correlation, 3rd ed.* Philadelphia: Lippincott-Raven, 1998.
10. Manning WJ, Pennell DJ. *Cardiovascular Magnetic Resonance.* New York: Churchill Livingstone, 2002.
11. Meszaros WT. *Cardiac Roentgenology: Plain Films and Angiocardiographic Findings.* Springfield: Charles C Thomas, 1969.
12. Miller SW. *Cardiac Radiology: The Requisites.* St. Louis: Mosby, 1996.
13. Mohan N. Kerr G. Aortitis. Curr Treat Options Cardiovasc Med 2002; 4:247-254.
14. Müller NL, Fraser RS, Colman NC, Paré PD. *Radiologic Diagnosis of Diseases of the Chest.* Philadelphia: Saunders, 2001.
15. Nakabayashi K, Kamiya Y, Nagasawa T. Aortitis syndrome associated with positive perinuclear antineutrophil cytoplasmic antibody: report of three cases. Int J Cardiol 2000; 75 Suppl 1:S89-94.
16. Nesi G, Anichini C, Pedemonte E, Tozzini S, Calamai G, Montesi GF, Gori F. Giant cell arteritis presenting with annuloaortic ectasia. Chest 2002; 121:1365-1367.
17. Parmley LF, Mattingly TW, Manion WC, Jahnke EJ Jr. Nonpenetrating traumatic injury of the aorta. Circulation 1953; 17:1086-1101.
18. Spindola-Franco H, Fish BG. *Radiology of the Heart: Cardiac Imaging in Infants, Children, and Adults.* New York: Springer-Verlag, 1985.
19. Woodring JH, Loh FK, Kryscio RJ. Mediastinal hemorrhage: an evaluation of radiographic manifestations. Radiology 1984; 151:15-21.
20. Woodring JH, Dillon ML. Radiographic manifestations of mediastinal hemorrhage from blunt chest trauma. Ann Thorac Surg 1984; 37:171-178.

21. Woodring JH, King JG. Determination of normal transverse mediastinal width and mediastinal-width to chest-width (M/C) ratio in control subjects: implications for subjects with aortic or brachiocephalic arterial injury. J Trauma 1989; 29:1268-1272.

22. Woodring JH, King JG. The potential effects of radiographic criteria to exclude aortography in patients with blunt chest trauma. J Thorac Cardiovasc Surg 1989; 97:456-460.

23. Woodring JH. The normal mediastinum in blunt traumatic rupture of the thoracic aorta and brachiocephalic arteries. J Emerg Med 1990; 8:467-476.

24. Woodring JH, Daniel TL, Bernardy MO. Chronic traumatic pseudoaneurysm of the thoracic aorta: recognition by computed tomography. J Ky Med Assoc 1984; 82:627-630.

Chapter 13.

CT of Coronary Artery Calcification

I. **Historical background**
   A. **Pathogenesis of coronary artery calcification (CAC)**
      1. CAC is a marker of coronary atherosclerosis.
      2. Monckeberg's medial sclerosis, which causes arterial calcification that is not associated with luminal narrowing, is extremely rare in the coronary arteries.
      3. CAC occurs almost exclusively in areas of intimal atherosclerosis and is frequently associated with luminal narrowing of the involved coronary arteries.
         a. Calcification occurs in lipid-laden plaque by several mechanisms.
            1) Calcification occasionally occurs in unruptured soft plaque.
            2) More importantly, calcification frequently occurs in complex plaque.
               a) Soft plaque is prone to rupture and develop ulceration.
               b) Ruptured or ulcerated plaque contains highly thrombogenic material, and clot forms at the site of rupture.
               c) Clots are partially lysed and partially incorporated into the plaque forming complex plaque.
               d) Over time, complex plaque begins to calcify.
               e) The volume of calcified plaque usually accounts for about 20% of the total atherosclerotic plaque burden.
         b. In individual lesions, the amount of calcification increases as the severity of stenosis increases.
      4. The left anterior descending coronary artery (LAD) and its branches (diagonal branches) are the most common site of CAC followed by the right coronary artery, circumflex coronary artery, and left main coronary artery.
   B. **Autopsy studies of CAC**
      1. The prevalence of CAC increases with increasing age.
      2. CAC is present in 53-80% of unselected patients over age 40, and is present in nearly 100% of patients over age 40 who have died from coronary atherosclerosis, regardless of age.
      3. The severity of CAC correlates with the severity of coronary arterial stenosis.
   C. **Clinical studies of CAC**
      1. The prevalence of CAC increases with increasing age.
      2. There are expected gender differences with women lagging behind men in development of CAC by about 10 years, reflecting the known lower incidence of coronary artery disease (CAD) in pre-menopausal women.

3. The amount of CAC correlates with the extent of coronary atherosclerosis.
4. There is a strong association between CAC and hyperlipidemia and hypercholesterolemia; however, there are many patients with coronary atherosclerosis and CAC who do not have hyperlipidemia or hypercholesterolemia.
5. The presence of hemodynamically significant stenosis correlates with the extent of CAC and the number of coronary arteries involved with CAC.
6. Patients who do not have CAC have a very low likelihood of significant coronary arterial stenosis and a low risk of developing symptomatic CAD.
7. The likelihood of symptomatic CAD increases as the amount of CAC increases.

D. Plain film identification of CAC
1. Plain films are relatively insensitive in detecting CAC; however, the identification of CAC on plain films has high specificity for the presence of significant CAD.
   a. CAC can be seen on plain films in up to 25% of patients clinically suspected of having CAD.
   b. CAC can be seen on plain films in up to 40% of patients with type II hyperlipoproteinemia.
   c. Unfortunately, most cases of CAC are not apparent on plain films.
   d. When CAC is visible on plain films, the patient has significant coronary atherosclerosis, a high likelihood of having at least one hemodynamically significant coronary arterial stenosis, and a high risk of developing symptomatic CAD.
2. CAC appears as linear or parallel tram-track calcification, or as circular calcification if seen end-on.
   a. Most of the CAC visible on plain films resides in the LAD. CAC in the other coronary arteries is rarely visible on plain films.
   b. Look in the coronary artery triangle on posteroanterior chest radiographs or near the location of the aortic valve on lateral views
   c. Calcification in the LAD on the lateral view often appears as tram-track calcification extending anteriorly from the region of the aortic root down along the anterior aspect of the cardiac silhouette toward the cardiac apex.

E. CAC identified by fluoroscopy
1. Fluoroscopy is superior to plain films in detecting CAC.
2. Both conventional image-intensifier fluoroscopy and digital fluoroscopy have been used to detect CAC.
3. CAC is identified by fluoroscopy in 25-66% of patients with clinical evidence of CAD, and in 56-92% of patients with hemodynamically significant stenosis on arteriography.
4. Symptomatic CAD is 2-3 times more likely in those with CAC identified by fluoroscopy than in those without fluoroscopic evidence of CAC.
5. Fluoroscopy, however, is no longer used to screen for CAC.

F. Identification of CAC on conventional CT

1. Conventional CT (non-gated, 1 cm contiguous slice thickness, 1-2 second scan time) is superior to fluoroscopy in identifying CAC.
2. Symptomatic CAD is 5.5 times more likely in patients under age 60 with CAC identified on conventional CT than in those without CAC.
3. Conventional CT, however, is inadequate for coronary artery calcium screening because it is limited in its ability to accurately quantify small areas of CAC due to motion artifact.

II. CAC screening utilizing fast CT techniques

A. Justification for using fast CT to screen for CAC

1. Coronary atherosclerosis exists for decades before the patient develops clinical evidence of CAD.
2. In one-third to one-half of cases the initial clinical manifestation of CAD is myocardial infarction (MI) or sudden death.
3. Lipid profiles fail to detect many patients destined to develop clinical CAD.
4. Risk factor intervention has been shown to be effective in decreasing clinical coronary events.
5. Detection of CAC by CT can identify individuals at high-risk for clinical CAD, that would not be identified otherwise, allowing early risk factor intervention prior to the development of a clinical coronary event.

B. Single-detector spiral CT

1. Like conventional CT, single detector spiral CT is considered inadequate for coronary artery calcium screening because of motion artifact.

C. Electron beam CT (EBCT)

1. EBCT is the gold standard for CAC screening.
2. Technique
   a. Contiguous, gapless axial images are obtained from the main carina through the apex of the left ventricle.
   b. 3 mm beam collimation, 3 mm table increments
   c. 130 kVp
   d. 630 mA
   e. ECG-gated single section mode, 100 millisecond acquisitions triggered at 80% of the R-R interval
   f. Single breath-hold
   g. Image reconstruction using 512 x 512 matrix, sharp reconstruction algorithm, and 26-30 cm display field of view
   h. All high attenuation foci within the coronary arteries with 2 or more contiguous pixels with an attenuation of 130 HU or greater are manually encircled with the calcium score being generated by proprietary software.
3. There is a strong correlation between the EBCT generated calcium score and actual histologic calcium measurements, and there is a strong correlation between the amount of CAC identified by EBCT in a given area and the histologic assessment of plaque severity in that area.

4. However, since the volume of calcified plaque only accounts for about 20% of the total volume of atherosclerotic plaque, EBCT underestimates total atherosclerotic plaque volume by 80%.
5. The identification of CAC by EBCT is 95-100% sensitive for detecting coronary atherosclerosis, and correspondingly is 95-100% sensitive for detecting hemodynamically significant coronary arterial stenosis. The specificity of CAC identified by EBCT for the presence of coronary atherosclerosis is likewise very high (essentially 100%); however, the specificity for the presence of a hemodynamically significant coronary arterial stenosis is lower, approximately 50%. Still, those patients who have CAC, but who do not have a hemodynamically significant stenosis, have nonobstructive coronary atherosclerosis and are at an increased risk for developing obstructive disease and symptomatic CAD.
6. A calcium scoring system is used to estimate the risk of developing symptomatic CAD.
7. As the quantity of CAC increases the likelihood of symptomatic CAD increases. A calcium score >75 increases the risk of MI or sudden death 6 times and a calcium score >160 increases the risk of MI or sudden death 35 times.
8. CAC is better than other risk factors in predicting such coronary events.
9. Although large, diffuse deposits of CAC are much more likely to be associated with hemodynamically significant luminal stenosis than small focal deposits, the location and extent of CAC cannot predict the exact location and severity of stenosis.

D. **Multi-detector helical CT (HCT)**
1. Is more widely available than EBCT
2. Technique
    a. Volumetric images are obtained from the main carina through the apex of the left ventricle.
    b. 2.5 mm collimation width x 4 detectors
    c. Pitch 1.0
    d. 0.8 sec rotational speed
    e. 120 kVp
    f. 200 mA
    g. Single breath-hold
    h. Retrospective ECG gating algorithm which provides 133 msec temporal resolution at 0.8 sec rotational speed
    i. The end of the temporal window is set at 80% of the R-R interval for multisector reconstructions.
    j. Images are reconstructed using 512 x 512 matrix, sharp reconstruction algorithm, and 26 cm display field of view.
    k. A threshold of 90 HU or 130 HU can be used for identification of CAC.
3. Compared to EBCT, images of the coronary arteries obtained by HCT with retrospective ECG gating have a higher signal-to-noise ratio and decreased motion artifact.

4. Calcium scores generated by HCT with retrospective ECG gating correlate well with calcium scores generated by EBCT. Results with 130 HU correlate better with EBCT, but 90 HU may detect CAC missed by EBCT.

E. **Coronary artery calcium scoring system (Agatston-Janowitz)**
   1. Calcium score 0
      a. No identifiable calcified atherosclerotic plaque
      b. Does not completely rule out presence of noncalcified soft plaque
      c. Very low risk of CAD
      d. Negative predictive value is >95% for absence of hemodynamically significant coronary stenosis.
      e. Adherence to general guidelines on diet and exercise is stressed.
   2. Calcium score 1-10
      a. Minimal plaque burden
      b. Significant obstructive CAD unlikely
      c. Adherence to general guidelines on diet and exercise is stressed.
   3. Calcium score 11-100
      a. Mild plaque burden
      b. Mild coronary arterial stenosis likely
      c. Recommend daily aspirin prophylaxis and adherence to National Cholesterol Education Program (NCEP) guidelines for cholesterol lowering
   4. Calcium score 101-400
      a. Moderate plaque burden
      b. Moderate coronary arterial stenosis likely
      c. High risk of developing symptomatic CAD
      d. Institute risk factor modification including daily aspirin prophylaxis, strict adherence to NCEP guidelines for cholesterol lowering, use of statin medication to reduce LDL cholesterol to <100 mg/dL, and institution of an appropriate exercise program.
      e. Clinical follow-up is necessary.
   5. Calcium score >400
      a. Extensive plaque burden
      b. Very high risk of developing symptomatic CAD
      c. High likelihood of at least one hemodynamically significant coronary arterial stenosis.
      d. Institute risk factor modification including daily aspirin prophylaxis, strict adherence to NCEP guidelines for cholesterol lowering, use of statin medication to reduce LDL cholesterol to <100 mg/dL, and institution of an appropriate exercise program.
      e. Perform non-invasive stress testing and consider coronary arteriography if positive.

F. **Coronary artery calcium scoring in symptomatic patients**
   1. CT screening for CAC has little if any role in patients who are already known to have CAD.

    2. CT screening for CAC, however, can be very helpful in patients with chest pain in whom CAD has not been documented.
       a. A calcium score of 0 essentially excludes the presence of CAD.
       b. Presence of CAC in a patient with chest pain suggests that the pain is related to CAD.

G. Coronary artery calcium scoring in asymptomatic patients
    1. Absence of CAC indicates absence of significant coronary atherosclerosis and a very low risk of developing symptomatic CAD.
    2. Presence of CAC identifies presence of coronary atherosclerosis.
    3. Results of calcium scoring can be used to guide risk factor intervention, and if >400, the need for stress testing and possible coronary arteriography.

References and suggested additional reading

1. Agatston AS, Janowitz WR, Hildner FJ, Zusmer NR, Viamonte M Jr, Detrano R. Quantification of coronary artery calcium using ultrafast computed tomography. J Am Coll Cardiol 1990; 15:827-832.
2. Bartel AG, Chen JT, Peter RH, Behar VS, Kong Y, Lester RG. The significance of coronary calcification detected by fluoroscopy: a report of 360 patients. Circulation 1974; 49:1247-1253.
3. Carr JJ. Coronary calcium: the case for helical computed tomography. J Thorac Imaging 2001; 16:16-24.
4. Dinsmore RE, Lees RS. Vascular calcification in types II and IV hyperlipoproteinemia: radiographic appearance and clinical significance. AJR 1985; 144:895-899.
5. Goldin JG, Yoon HC, Greaser LE III, Heinze SB, McNitt-Gray MM, Brown MS, Sayre JW, Emerick AM, Aberle DR. Spiral versus electron-beam CT for coronary artery calcium scoring. Radiology 2001; 221:213-221.
6. Hamby RI, Tabrah F, Wisoff BG, Hartstein ML. Coronary artery calcification: clinical implications and angiographic correlates. Am Heart J 1974; 87:565-570.
7. Horiguchi J, Nakanishi T, Ito K. Quantification of coronary artery calcium using multidetector CT and a retrospective ECG-gating reconstruction algorithm. AJR 2001; 177:1429-1435.
8. Janowitz WR. CT imaging of coronary artery calcium as an indicator of Atherosclerotic disease: an overview. J Thorac Imaging 2001; 16:2-7.
9. Kelley MJ, Huang EK, Langou RA. Correlation of fluoroscopically detected coronary artery calcification with exercise stress testing in asymptomatic men. Radiology 1978; 129:1-6.
10. Langou RA, Kelley MJ, Huang EK, Cohen LS. Predictive accuracy of coronary artery calcification and positive exercise test in asymptomatic non-hyperlipidemic men for coronary artery disease (Ab). Am J Cardiol 1980; 45:400.
11. Sousa AS Jr, Bream PR, Elliott LP. Chest film detection of coronary artery calcification: the value of the CAC triangle. Radiology 1978; 129:7-10.

12. Thompson BH, Stanford W. Imaging of coronary calcium: a case for electron beam computed tomography. J Thorac Imaging 2001; 16:8-15.
13. Woodring JH, West JW. Coronary artery calcification identified by CT in patients over 40 years of age. Australas Radiol 1989; 33:79-83.

Chapter 14.

Cardiac MR

I. **Current clinical uses of cardiac MR**
   A. **Cardiac MR is useful in the evaluation of a number of cardiac abnormalities including the following:**
      1. Pericardial disease
      2. Aortic dissection
      3. Acquired heart disease
      4. Cardiac tumors
      5. Myocardial disease
      6. Coronary atherosclerosis and ischemic heart disease
      7. Congenital heart disease
   B. **Advantages of MR**
      1. MR is a noninvasive, accurate, and reproducible means of assessing cardiac function and anatomy.
      2. There is high intrinsic contrast between flowing blood and stationary tissue.
      3. MR utilizes a tomographic format with high resolution.
      4. There is a large field of view, which allows the heart, pulmonary arteries, thoracic aorta, and great veins to be readily appreciated.
      5. Contrast material is not absolutely necessary.
   C. **Relative disadvantages of MR**
      1. MR is more costly and time-consuming than echocardiography.
      2. There is relatively poor temporal resolution.
      3. There can be variable intraluminal signal, which is most apparent on spin echo imaging.
      4. Physiologic data often must be calculated by hand.
      5. Fast gradient-echo imaging and newer software packages are largely eliminating these disadvantages.

II. **Technical considerations**
   A. **Synchronized data acquisition**
      1. Diagnostic images require synchronized data acquisition. This is accomplished by ECG triggering which is based upon R-wave detection.
      2. ECG triggering is obscured by an induced electric potential caused by flowing blood in the magnetic field.
         a. In the supine position the left ventricle (LV) and ascending aorta conduct blood nearly orthogonal to the magnetic field of the imager.
         b. This potential may increase the amplitude of the ECG T-wave confusing the R-wave detection of the imager.
      3. This may be minimized by careful lead placement.

    a. ECG surface electrodes should be distributed over the left anterior chest in a standard manner across the precordium.

    b. Surface electrodes should be distributed within 15 cm of each other.

    c. Leads should pass directly over the patient's left shoulder without forming loops.

    d. Be prepared to change the arrangement of lead attachments to the surface electrodes. The combination of electrode placement and lead attachment that produces the tallest R-waves with the least noisy background is the best lead selection.

B. Types of acquisitions

1. Spin echo sequences

    a. Spin echo sequences remain the mainstay of cardiac MR imaging.

    b. First, spins are excited by a $90°$ radio frequency (RF) pulse in the presence of a slice-selection gradient. Spin magnetization is then refocused using a $180°$ RF pulse that is also slice-selective. Following this the MR signal is sampled with a readout gradient, and the process is repeated with varying degrees of phase encoding, ranging from 64 to 512 steps for cardiac imaging.

    c. Repetition time (TR) is set by the length of the patient's ECG R-R interval and is, therefore, variable. Typical values may range from 550-900 msec.

    d. Echo time (TE) is typically 20 msec; for dual echo images echo times of 20 and 40 msec are used.

    e. Raw data collected is known as "k-space."

    f. Spin echo sequences produce static "black blood" images. This effect is the result of the $90°$-$180°$ pulse pair.

        1) The $90°$ pulse sets up a window of excitation and only spins remaining in this window at the time of the $180°$ pulse are refocused and detected.

        2) Blood flowing rapidly through the chosen image plane causes the excited spins in a blood vessel or cardiac chamber to leave the image plane and be replaced by unexcited spins, which are not refocused and detected.

        3) This results in signal loss within the lumen of the vessel or cardiac chamber.

        4) Because arterial blood flow is pulsatile, spin echo images obtained during diastole, when blood flow may be very slow, may have medium or high signal intensity.

    g. Spin echo sequences are good for anatomical detail and are used for evaluating morphologic abnormalities of the mediastinum, pericardium, heart, and great vessels.

2. Cine gradient-echo MR imaging

    a. A single slice-selective RF pulse of flip angle α (where α is usually $<90°$) is followed in rapid succession by a phase-encoding gradient and a two-part readout gradient, in which reversal of the gradient polarity takes the place of the $180°$ spin echo refocusing pulse in generating a signal echo. Acquisition of the entire k-space signal occurs by repetition of the α pulse with different phase-encoding gradient steps. Gradient-echo acquisition is generally faster than spin echo acquisition because less time is required between spin excitation and signal detection.

b. Typically performed with a TR of 20-30 msec, TE of 5-12 msec, and a flip angle of 30° and takes only 2-5 minutes to perform.

c. Sixteen frames are usually sufficient to evaluate the entire cardiac cycle and are displayed in a cine loop allowing a dynamic approach.

d. Cine gradient-echo sequences produce "white blood" images where blood has bright signal intensity. Unlike spin echo sequences, there is no selective window of detection to match the excitation window. Because of rapid repetition of the α pulses, spins do not have time to relax back to their unexcited state via T1 relaxation. This results in partial saturation of spins in the slice. Fresh blood flowing into the slice of interest has not been saturated and therefore contributes a greater signal than surrounding stationary tissue that has undergone multiple excitations.

e. Turbulent blood flow causes dephasing of spins within a voxel and results in a dark area of signal loss or flow void.

  1) This flow void is encountered with valvular stenosis and regurgitation and intracardiac shunts. Its appearance depends on technical factors including window width and level, flip angle, and particularly TE.

  2) With long-TE sequences (12 msec) the flow void is well demonstrated.

  3) With short-TE sequences (<7 msec) the flow void tends to disappear.

f. Segmentation of k-space (phase-encoding grouping) is a method of data collection that can decrease acquisition time for cine gradient-echo imaging by a factor of 2 or 3 making it possible to use long-TE sequences to evaluate abnormal flow patterns while maintaining short acquisition times.

g. Cine gradient-echo sequences are particularly useful for evaluation of cardiac function and identifying abnormal flow patterns seen in valvular stenosis and regurgitation.

3. Fast gradient-echo pulse sequences
  a. TR/TE = 4-8 msec/2-4 msec
  b. Can be used to obtain cine MR images during a single breath-hold thus eliminating respiratory artifacts
  c. These sequences are not appropriate for evaluating flow abnormalities because of the short TE.
  d. Ideal for evaluating cardiac wall motion abnormalities, calculating ejection fraction, and determining cardiac mass
4. Flow-sensitive MR imaging (velocity-encoded cine or VEC MR imaging)

    a. Based on the principle that the phase of flowing spins relative to stationary spins along a magnetic gradient changes in direct proportion to flow velocity

    b. Following a 90° RF excitation pulse, a bipolar velocity-encoding gradient (G_{vel}) is applied. The first portion of the bipolar gradient establishes a phase shift, and the second portion, which is of reversed polarity, refocuses that phase shift.

    c. Stationary spins acquire equal and opposite phases in the two gradients and are left with no net phase at the end of the sequence. Moving spins will encounter different regions of the two gradients and will be left with a net phase that is proportional to their velocity in the direction of the velocity-encoding gradient. Phase-sensitive data collection and reconstruction results in an image whose magnitude describes the signal intensity and whose phase describes the velocity at each pixel.

      1) The phase shift is displayed as variations in pixel signal intensity on the phase map image.

        a) Stationary tissue appears gray on this image, whereas flow in a positive direction along the velocity-encoding gradient will appear bright and flow in a negative direction will appear dark.

        b) It is therefore possible to distinguish antegrade from retrograde flow.

        c) Allows quantification of blood velocity profiles at different times during the cardiac cycle

        d) Allows quantitative assessment of valvular stenosis and regurgitation

      2) Magnitude images can be reconstructed for anatomic detail.

    d. Because of the cyclic nature of phase, aliasing may appear if more than one cycle of phase shift occurs. This can be avoided by keeping the velocity threshold set to maintain a phase shift of less than 180°.

    e. While long-TE sequences are necessary for visualizing signal loss due to turbulent blood flow on cine gradient-echo sequences, short-TE sequences are necessary for accurately measuring the velocity of turbulent jets due to valvular stenosis or regurgitation on VEC MR.

    f. Misalignment between true and measured flow influences the measurement of flow velocity as determined by the equation

$$V_{meas} = V_{true} \, (\cos \theta)$$

where V_{meas} is the measured velocity, V_{true} is the true velocity, and θ is the angle of misalignment between flow encoding and flow direction. In most cases the angle of misalignment is small and the error in flow velocity is also small. A misalignment of 20° produces an error of only 6%.

  5. Echoplanar MR imaging (EPI)

    a. Single-shot EPI

1) Fast gradient-echo techniques require one RF pulse for each k-space line. With single-shot EPI multiple k-space lines are acquired after a single RF pulse. A rapidly oscillating readout gradient generates a series of gradient-echo signals each of which is separately phase-encoded by a gradient blip in the phase-encoding direction.
2) Single-shot EPI is the fastest of all the very fast MR techniques. All imaging data is usually acquired in 30-100 msec permitting true real-time MR imaging.
3) Single-shot EPI can be used to evaluate ventricular size and function and to evaluate the coronary arteries.
  b. Multishot EPI is a variant of single-shot EPI in which several pulses are used to obtain an interleaved series of data. Although it can be obtained in a single breath-hold, it requires several cardiac cycles and is not a real-time technique. Multishot EPI can be used to evaluate ventricular size and function and to evaluate the coronary arteries.

C. Routine imaging planes
1. Axial plane
   a. In the axial plane the anatomy displayed is similar to that on CT.
   b. Useful in evaluating congenital and acquired heart disease
2. Sagittal and coronal planes
   a. Useful in evaluating the aorta, main pulmonary artery (MPA), and pulmonary arteries
3. Right anterior oblique sagittal plane
   a. Parallel to interventricular septum
   b. Provides images parallel to the vertical long-axis of the LV and right ventricle (RV)
4. Left anterior oblique sagittal plane
   a. Parallel to the atrioventricular ring and the aortic arch
   b. Good for demonstrating right ventricular outflow tract (RVOT), MPA, and thoracic aorta
5. Horizontal short-axis view
   a. The LV myocardium is displayed as a series of doughnut-shaped rings.
   b. Good for evaluation of LV myocardial thickening and radial contraction
6. Horizontal long-axis view
   a. Images are obtained parallel to the long-axis of the LV.
   b. Good for evaluation of the LV wall and septum and the aortic and mitral valves
7. Four-chamber view
   a. Images are parallel to the LV long-axis and orthogonal to its vertical short-axis. The images are tilted superoinferiorly such that as one moves from the top to the bottom of the image one is also moving ventrally as well as inferiorly.
   b. Displays all four chambers simultaneously and gives a good view of the atrial septum, the membranous portion of the ventricular septum,

as well as its muscular portion, and the tricuspid, mitral, and aortic valves
8. Two-chamber view
   a. Images are obtained parallel to a plane passing from the middle of the mitral valve to the LV apex.
   b. Useful for evaluating the left atrium (LA), mitral valve, and LV

III. Pericardial disease
A. Normal pericardium
1. 1-2 mm thickness typical; 3 mm also considered normal
2. Low signal intensity on spin echo images
B. Pericardial effusion
1. Simple transudative effusions typically have low signal intensity on T1-weighted images and high signal intensity on T2-weighted images.
2. Slow motion of pericardial fluid may occasionally cause medium signal intensity on T1-weighted images.
3. Hemorrhagic or exudative effusions typically have areas of medium or high signal intensity on T1-weighted images.
4. Pericardial effusion has high signal intensity similar to blood on cine gradient-echo sequences because of motion of the fluid in the pericardial cavity.
C. Pericardial tamponade
1. There is usually a large amount of pericardial effusion.
2. Increased pressure in the pericardial sac causes compression and flattening of the atria resulting in diminished atrial filling.
D. Constrictive pericarditis
1. Antecedent, often clinically unapparent, viral infection is thought to be the most common cause of constrictive pericarditis today. Other causes include tuberculosis, tumor infiltration, prior mediastinal irradiation, trauma, or cardiac surgery.
2. Thickened pericardium 4 mm or more, with medium signal intensity on T1-weighted images, is strongly suggestive of constrictive pericarditis (accuracy, 93%).
   a. When spin echo sequences suggest pericardial thickening it is advisable to perform cine gradient-echo imaging as well. Because of low signal intensity of pericardial effusion on spin echo sequences it may be difficult to distinguish small amounts of effusion from pericardial thickening.
   b. On cine gradient-echo images pericardial effusion will have high signal intensity similar to blood, whereas pericardial thickening will have medium signal intensity similar to soft tissue.
3. Associated findings
   a. Signs of systemic venous hypertension including distention of the inferior vena cava (IVC), ascites, and pleural effusions
   b. Diminished size of the ventricular cavities
   c. May also have pericardial effusion (effusive-constrictive pericarditis)

    d. Myocardial fibrosis with thin-walled, poorly-contractile LV

       1) Patients with associated myocardial fibrosis do not respond well to surgical stripping/pericardiectomy.

4. Unlike CT, MR is not good for demonstrating pericardial calcification. Because of this, CT is often preferred over MR for initial evaluation of constrictive pericarditis.

5. In a patient clinically suspected of having constrictive pericarditis, the demonstration of pericardial thickening strongly supports the diagnosis. However, there can be pericardial thickening without constriction, particularly in patients with prior cardiac surgery or postpericardiotomy syndrome, and there can be constrictive pericarditis without pericardial thickening.

6. Demonstration of pericardial thickening on MR is also useful in excluding restrictive cardiomyopathy as a cause of the patient's symptoms.

E. Pericardial tumor

1. Common malignant tumors involving the pericardium include metastatic breast and lung carcinoma, melanoma, and lymphoma. Involvement may be by direct extension or hematogenous dissemination. Mesothelioma, which is the most common primary pericardial tumor, is rare.

2. MR may show focal nodular thickening of the pericardium or hemorrhagic effusion with high signal intensity on T1-weighted images.

3. Mesotheliomas and other malignant pericardial tumors usually enhance with gadolinium-DTPA (Gd-DTPA).

F. Congenital pericardial abnormalities

1. Congenital absence of the pericardium

    a. Thought to be due to premature obliteration of the embryonic duct of Cuvier leading to incomplete formation of the parietal pericardium

    b. Partial absence of the pericardium is more common than complete absence of the pericardium.

    c. The absent portion of the pericardium is typically on the left side.

    d. Imaging findings

       1) Chest radiography typically demonstrates significant leftward displacement of the heart, posterior rotation of the cardiac apex, a prominent MPA, and interposition of lung between the inferior margin of the heart and the left hemidiaphragm. Occasionally the left atrial appendage (LAA) may be prominent.

       2) This cardiac configuration is so characteristic that other imaging modalities such as CT or MR are usually not necessary. In fact, it is not uncommon for a portion of the normal pericardium to be visualized poorly or not at all on CT and MR. Therefore, failure to visualize part of the pericardium on CT or MR is not indicative of congenital absence of the pericardium.

       3) The characteristic finding of congenital absence of the pericardium on CT or MR is the finding of aerated lung situated between the aorta and MPA (most common) or between the MPA and LAA. Normally the left pericardium keeps lung out of these recesses.

4) The presence of a circumscribed crease in the LV wall suggests associated cardiac strangulation (rare).

2. Pericardial cyst
   a. Seen as a round or oval structure often contiguous with normal pericardium
      1) A typical benign pericardial cyst should have a thin, almost imperceptible wall.
      2) If the cyst wall is thick or nodular suspect a necrotic tumor such as a malignant teratoma instead.
   b. Typically has low signal intensity on T1-weighted spin echo images and high signal intensity on T2-weighted images and cine gradient-echo images
   c. If the cyst has high protein content or is complicated by hemorrhage it may have medium or high signal intensity on T1-weighted images.

IV. Aortic dissection
A. Advantages of MR
1. MR has both high sensitivity and specificity for aortic dissection and is superior to both aortography and CT in the diagnosis of aortic dissection. MR has comparable sensitivity to transesophageal echocardiography (TEE) in the detection of aortic dissection (96-100% sensitivity for both modalities), but is significantly more specific for aortic dissection than TEE (98-100% versus 85%-97%, respectively).
2. MR allows detection of complications of aortic dissection including involvement of great vessels and coronary arteries, aortic regurgitation, pericardial tamponade, and pleural effusion.

B. Disadvantages of MR
1. MR cannot demonstrate displaced intimal calcification.
2. Due to relatively slow acquisition times and difficulty in monitoring critically ill patients, MR is used mainly in stable patients.

C. MR findings
1. On spin echo sequences the intimal flap appears as a line of medium signal intensity separating signal void within the true and false lumen, or an interface between different signal intensities within the true and false lumens.
   a. The true lumen usually has higher flow velocity and a complete signal void. The false lumen may demonstrate medium signal intensity due to slower flow.
   b. Small linear structures projecting from the wall of the false lumen to the intimal flap (aortic "cobwebs") may be seen in 20% of cases and are a reliable marker of the false lumen.
   c. Entry sites are identified as focal interruptions in the intimal flap.
2. Cine gradient-echo sequences show the intimal flap as a line of low-to-medium signal intensity separating higher signal intensity flowing blood in the true and false lumens.
   a. The true lumen usually has high signal intensity because of rapid flow.

     b. The false lumen tends to have medium signal intensity because of slower flow or thrombosis.
       1) Thrombus is seen as areas of low-to-medium signal intensity that do not change significantly during the cardiac cycle.
       2) Slow flow is characterized by low-to-medium signal intensity that changes during different portions of the cardiac cycle.
   3. Differentiation of dissection with completely thrombosed false lumen from atherosclerotic aneurysm with mural thrombus may be difficult.
     a. MR cannot demonstrate displaced intimal calcification.
     b. Thickening of the aortic wall may be only sign of aortic dissection in up to 14% of cases.
       1) Thrombosed false lumen may only cause minimal compression of true lumen.
       2) T1-weighted images show hyperintense foci consistent with subacute intramural hemorrhage in the aortic wall in 80% of these cases. This is consistent with early intramural hemorrhage occurring before the development of intimal rupture and blood flow in a false lumen.

V. Acquired heart disease
A. Mitral valve disease
   1. Mitral stenosis is usually caused by rheumatic heart disease; mitral regurgitation may be secondary to ischemic heart disease, myocardial infarction, or endocarditis.
   2. In mitral stenosis, cine gradient-echo sequences show left atrial enlargement (LAE), thickening and bulging of the mitral valve, and a signal void in the LV (jet phenomenon) distal to the mitral valve in diastole.
   3. In mitral regurgitation, cine gradient-echo sequences show a systolic signal void in the LA above the mitral valve. The LA and LV are typically enlarged.
B. Aortic valve disease
   1. Aortic stenosis may be due to a congenital bicuspid valve, valve degeneration, or rheumatic disease; aortic regurgitation may be due to rheumatic heart disease, annuloaortic ectasia, aortic dissection, hypertension, or myocardial infarction.
   2. Cine gradient-echo sequences can demonstrate whether the aortic valve has 2 or 3 cusps.
   3. In aortic stenosis, cine gradient-echo sequences show thickening and bulging of the aortic valve with a systolic signal void in the ascending aorta above the aortic valve. There is usually post-stenotic dilatation of the ascending aorta and LV hypertrophy. If the aortic stenosis is long-standing, there may be LV dilatation.
   4. In aortic regurgitation, cine gradient-echo images show a diastolic signal void in the LV below the aortic valve. The LV may be enlarged. In

annuloaortic ectasia there is uniform dilatation of the aortic annulus and ascending aorta.

C. Tricuspid valve disease
1. Tricuspid stenosis is rare. It is usually due to rheumatic heart disease or endocarditis. In tricuspid stenosis, cine gradient-echo images show enlargement of the right atrium (RA), thickening and bulging of the tricuspid valve, and a diastolic signal void in the right ventricle (RV).
2. Tricuspid regurgitation is the most common form or right-sided valvular heart disease. Tricuspid regurgitation can be secondary to pulmonary arterial hypertension (PAH), mitral stenosis, endocarditis, RV infarction, and surgery for congenital heart disease.
   a. PAH is the most common cause of tricuspid regurgitation.
   b. In tricuspid regurgitation, cine gradient-echo sequences show enlargement of the RA and RV and a systolic signal void in the RA above the tricuspid valve.

D. Multivalvular heart disease
1. Mitral stenosis with aortic regurgitation
   a. The dominant signs are usually those of mitral stenosis.
   b. LV filling from the LA is impaired. There is LAE, evidence of pulmonary venous hypertension, and often evidence of PAH as well.
   c. Despite the presence of aortic regurgitation, the LV is usually not greatly enlarged.
2. Mitral stenosis with tricuspid regurgitation
   a. The findings are predominantly those of mitral stenosis with added enlargement of the right heart.
   b. In these cases tricuspid regurgitation is usually secondary to long-standing PAH caused by the mitral stenosis.
3. Combined mitral and aortic regurgitation
   a. The combination of mitral and aortic regurgitation results in marked enlargement of both the LA and LV.
   b. Signs of pulmonary venous hypertension may be present; however, signs of PAH are usually absent.
4. Combined mitral and aortic stenosis
   a. The dominant signs are those of mitral stenosis.
   b. The signs of aortic stenosis are usually mild since obstruction of blood flow through the mitral valve minimizes the sequelae of the aortic stenosis.
5. Aortic stenosis with mitral regurgitation
   a. Usually caused by rheumatic heart disease, but can be caused by the coexistence of a stenotic bicuspid aortic valve and mitral valve prolapse with mitral regurgitation
   b. Dilatation of the LV secondary to aortic stenosis contributes to the severity of mitral regurgitation and can cause marked enlargement of both the LA and LV.
6. Combined valvular disease involving the tricuspid, mitral, and aortic valves is almost exclusively due to rheumatic heart disease.

E. MR quantitation of valvular disease
1. Assessment of valvular stenosis
   a. Evaluation of the flow jet and associated findings
      1) Cine gradient-echo sequences can be used to evaluate the size and extent of the abnormal flow jet.
         a) The jet can be outlined manually as a region of interest and the area of the signal void can be calculated by the imaging software.
         b) This technique provides only a semiquantitative index of the severity of the stenosis.
      2) In addition, thickening and bulging of the valve leaflets, reduced valve motion, chamber enlargement and wall thickening can be identified.
   b. Quantification of transvalvular pressure gradient and valve area with VEC MR imaging
      1) The pressure gradient across the stenotic valve (P_{grad}) can be determined by measuring the velocity of the turbulent jet of blood coming through the stenotic valve.
      2) Typically velocity measurements are made in planes perpendicular (through-plane velocity measurement) and parallel (in-plane velocity measurement) to the direction of flow.
      3) The relationship between the velocity of the jet and the difference in pressure on either side of the stenotic valve can be approximated by the modified Bernoulli equation, which in its simplest form is

$$P_1 - P_2 = 4(V_2^2 - V_1^2)$$

where P_1 is the pressure proximal to the stenotic valve, P_2 is the pressure distal to the stenotic valve, V_1 is the velocity proximal to the stenotic valve, and V_2 is the velocity distal to the stenotic valve. When velocity is expressed in m/sec, the pressure gradient is in mm Hg. In most clinical situations, V_1 is rarely greater than 1 m/sec and is usually disregarded. The fully modified equation is

$$P_{grad} = 4V_2^2$$

      4) A short TE must be used in order to recover signal from the jet core.
         a) A TE of 6-8 msec is used for low-velocity jets such as may occur at the atrioventricular level (tricuspid and mitral valves).
         b) A TE of 4-6 msec is used for medium velocity jets such as may occur with mild to moderate obstruction of the outflow valves (pulmonic or aortic valves) or surgical conduits.
         c) A TE of 3 msec or less is used for high velocity jets (4-6 m/sec) such as may occur with severe obstruction of the outflow valves or surgical conduits.

5) The valve area can also be calculated with the following formula

$$A_{valve} = (A_{OT} \times V_{OT}) / V_{jet}$$

where A_{valve} is the valve area, A_{OT} is the area of the outflow tract of the valve, V_{OT} is the maximum velocity in the outflow tract, and V_{jet} is the maximum velocity measured in the turbulent jet.
The area of the outflow tract is determined by the following formula

$$A_{OT} = \pi r^2$$

where r, the radius of the outflow tract, is one-half the diameter, or width, of the outflow tract.

6) Grading of severity of aortic stenosis
   a) Grading of aortic valve area (AVA) in aortic stenosis
      i) AVA 2.6-3.6 cm^2 — normal
      ii) AVA 1.5-2 cm^2 — mild stenosis
      iii) AVA 1.1-1.5 cm^2 — moderate stenosis
      iv) AVA <1 cm^2 — severe stenosis
   b) Grading of transvalvular pressure gradient in aortic stenosis at rest
      i) Pressure gradient 20-40 mm Hg — mild stenosis
      ii) Pressure gradient 40-50 mm Hg — moderate stenosis
      iii) Pressure gradient >50 mm Hg — severe stenosis

7) Grading of severity of mitral stenosis
   a) Grading of mitral valve area (MVA) in mitral stenosis
      i) MVA 4-6 cm^2 — normal
      ii) MVA 1.5-2.5 cm^2 — mild stenosis
      iii) MVA 1-1.4 cm^2 — moderate stenosis
      iv) MVA <1 cm^2 — severe stenosis
   b) Grading of transvalvular pressure gradient in mitral stenosis at rest
      i) Pressure gradient 4-6 mm Hg — mild stenosis
      ii) Pressure gradient 7-9 mm Hg — moderate stenosis
      iii) Pressure gradient 10 mm Hg or more — severe stenosis

8) Because the transvalvular pressure gradient can vary greatly for any given valvular area depending upon such factors as heart rate and cardiac output, which in turn affect the amount of blood flowing across the stenotic valve, the valve area is generally considered to be a better indicator of the severity of valvular stenosis than the transvalvular pressure gradient. For example, a patient with an MVA <1 cm^2 may have a transmitral pressure gradient of 10 mm Hg at rest that could increase to 25 mm Hg or more with only slight exertion.

2. Assessment of valvular regurgitation
   a. Cine gradient-echo sequences can be used to evaluate the size and extent of the abnormal flow jet.
      1) The jet can be outlined manually as a region of interest and the area of the signal void can be calculated by the imaging software.
      2) This technique provides only a semiquantitative index of the severity of the valvular regurgitation.
   b. Calculation of the regurgitant fraction with ventricular volumetric measurements.
      1) Accurate estimates of ventricular volume and ventricular mass can be obtained from cine gradient-echo images using the Simpson method.
         a) The RV and LV are imaged from base to apex in short-axis projection and a series of end-diastolic and a series of end-systolic images are generated for each level.
         b) The interventricular septum is considered part of the LV.
         c) LV end-diastolic volume and LV mass are calculated as follows.
            i) On the series of end-diastolic images the epicardial and endocardial borders of the LV are traced on each image. The anterior edge of the interventricular septum is considered part of the "epicardial" margin of the LV.
            ii) The areas of the LV cavity on each end-diastolic image are added together and multiplied by the slice thickness to provide the end-diastolic volume of the LV.
            iii) The areas of the LV encompassed by the traced epicardial margin of the LV on each end-diastolic image are added together and multiplied by the slice thickness to provide a total LV volume. Subtracting the LV end-diastolic volume from this volume provides the volume of tissue in the LV wall.
            iv) LV mass is calculated as the volume of tissue in the LV wall multiplied by an assumed density of 1.05 gm/cm^3.
         d) LV end-systolic volume is similarly calculated. The endocardial surface of the LV is traced on each of the end-systolic images. The areas of the LV cavity on each of the end-systolic images are added together and multiplied by the slice thickness to provide the end-systolic volume of the LV.
         e) RV end-systolic volume and end-diastolic volume are calculated in the same way.
         f) RV mass is based only on the volume of the RV free wall.
      2) A rough estimate of ventricular volume can be obtained using a simplified version of the Simpson method. On cine gradient-echo images the area of the ventricles is measured on two short-axis views and the length of the ventricular cavities is measured on a four-chamber view. Ventricular volume is defined as

$$V = L/2 \ (A_{MV} + 2/3 \ [A_{PM}])$$

where V is the ventricular volume, L is the length of the ventricle, A_{MV} is the area of the ventricle on the short-axis view obtained immediately below the mitral valve and A_{PM} is the area of the ventricle on the second short-axis view obtained at the level of the papillary muscles.

3) Normally the right ventricular stroke volume and the left ventricular stroke volume are equal.
4) The ventricular stroke volume is the difference between the end-diastolic volume and the end-systolic volume.
5) The difference in stroke volume between a regurgitant ventricle (the LV in mitral and aortic regurgitation) and a normal ventricle (the RV) is the regurgitant volume.
6) The regurgitant fraction is calculated by dividing the regurgitant volume by the stroke volume of the regurgitant ventricle.
   a) A regurgitant fraction of 15-20% indicates mild regurgitation.
   b) A regurgitant fraction of 20-40% indicates moderate regurgitation.
   c) A regurgitant fraction >40% indicates severe regurgitation.
7) Limitations of this method
   a) The method is more reliable for the LV than the RV.
   b) The method cannot be used in cases of multivalvular disease.
c. Quantification of valvular regurgitation with VEC MR imaging
   1) The cardiac cycle is divided into 16 equal time frames.
   2) On each of the 16 time frames on the magnitude study, the cross-sectional area is measured from a manually drawn region of interest around the pertinent vascular structure (aorta, pulmonary artery, mitral or tricuspid annulus).
   3) The product of the cross-sectional area times the average velocity in the region of interest gives the flow volume for each time frame on the magnitude study and the sum of all 16 flow volumes from the magnitude study provides the estimated flow volume for each heartbeat.
   4) This region of interest is then transferred from the magnitude image to the corresponding phase image for each time frame.
   5) Since antegrade flow on the phase study has bright signal and retrograde or regurgitant flow is dark, the flow volume of antegrade flow and the flow volume of regurgitant flow can be determined from the phase study.
      a) The stroke volume is the sum of all the flow volumes for the frames of the phase study demonstrating antegrade flow.
      b) The regurgitant volume is the sum of all the flow volumes for the frames of the phase study demonstrating retrograde flow.
      c) The regurgitant fraction is calculated by dividing the regurgitant volume by the stroke volume.

VI. Cardiac tumors
  A. Right atrial pseudotumor
    1. The crista terminalis, a normal muscular ridge along the posterolateral wall of the RA, can produce a triangular, sometimes rounded, projection that can be misinterpreted as a tumor.
    2. It is seen on both spin echo and cine gradient-echo sequences and has signal intensity similar to myocardium.
    3. The crista terminalis can be seen on MR in 59-90% of cases.
  B. Benign cardiac tumors
    1. 75% of cardiac tumors
    2. Myxoma
      a. Myxomas account for 50% of benign cardiac tumors.
      b. 70% occur between the third and sixth decades of life
      c. Location
        1) LA — 75%
        2) RA — 20%
        3) RV or LV — 5%
      d. Myxomas are characteristically attached to the atrial septum — 85% arise from the region of the fossa ovalis. Only a few arise from the posterior atrial wall or atrial appendage. They are typically polypoid and often pedunculated.
      e. Clinically myxomas usually present with either signs of systemic embolism or signs of obstruction of the mitral valve.
      f. MR findings
        1) T1-weighted spin echo
          a) Medium signal intensity
          b) May be heterogeneous due to calcification (low signal intensity) or hemorrhage (high signal intensity)
        2) T2-weighted spin echo
          a) Often higher signal intensity than myocardium
          b) May have low signal intensity if iron-laden
        3) Cine gradient-echo
          a) Low signal intensity compared to surrounding blood
          b) Motion of the mass is usually demonstrated and obstruction of the mitral valve during diastole may be demonstrated.
        4) Enhancement with Gd-DTPA
          a) After administration of Gd-DTPA the signal intensity increases.
        5) Attachment to the atrial septum is a characteristic finding.
      g. Myxomas can be difficult to distinguish from LA thrombus
        1) Fresh thrombi have higher signal intensity than myocardium on T1-weighted spin echo images and the contrast is further accentuated on T2-weighted spin echo images consistent with a high amount of hemoglobin.
        2) After 1-2 weeks paramagnetic compounds in the organizing thrombus (deoxyhemoglobin and methemoglobin) cause T1 and

T2 shortening, resulting in increased signal intensity on T1-weighted and decreased signal intensity on T2-weighted images. On cine gradient-echo sequences thrombus has the lowest signal intensity compared to surrounding cardiac structures and surrounding blood has the highest signal intensity.

3) Thrombus usually does not enhance with Gd-DTPA unless it is organized.

4) Thrombus is usually attached to the wall of the involved cardiac chamber, but is seldom attached to the atrial septum. Furthermore, thrombus does not move around on cine gradient-echo sequences. MR identification of a mobile intraatrial mass attached to the atrial septum strongly favors myxoma.

3. Lipoma and lipomatous hypertrophy
   a. True lipomas are the second most common benign cardiac tumor, accounting for 10% of cases.
      1) True lipomas are encapsulated and contain neoplastic cells.
      2) There is no sex or age predilection.
      3) Location
         a) Sub-endocardium — 50%
            i) Often arise from the atrial septum
            ii) Usually located in RA
         b) Sub-epicardium — 25%
            i) May become quite large
            ii) Can alter cardiac function
            iii) Can involve coronary arteries
         c) Myocardium — 25%
   b. Lipomatous hypertrophy of the atrial septum is more common than a true lipoma.
      1) Seen in females, the elderly, and patients who are obese
      2) Represents non-encapsulated hyperplasia of adipose tissue which may extend from the atrial septum to the lateral atrial wall, atrioventricular groove, free wall of the RV, and superior vena cava (SVC)
      3) Usually asymptomatic; however, there is often a history of atrial fibrillation
   c. MR findings
      1) Similar to subcutaneous fat, lipomas and lipomatous hypertrophy show high signal intensity on T1-weighted images and a slight decrease in signal intensity on T2-weighted images.
      2) Reduced or eliminated signal on fat-saturation images confirms the diagnosis.
   d. CT shows attenuation coefficients of about –100 HU.
4. Fibroma
   a. Congenital benign tumor usually discovered in children or occasionally young adults

    b. Fibromas arise in the myocardium, typically the LV anterior free wall or interventricular septum.

    c. Can cause blood flow obstruction

    d. Increased risk of sudden death secondary to arrhythmias

    e. MR findings

       1) T1-weighted spin echo

          a) Isointense to slightly hyperintense compared to myocardium

          b) May be heterogeneous due to calcification (low signal intensity) or hemorrhage (high signal intensity)

       2) T2-weighted spin echo

          a) Decreased in signal intensity compared to surrounding myocardium

            i) Due to fibrotic nature and presence of calcification due to necrosis

            ii) CT can easily confirm the presence of calcification in fibromas.

       3) Cine gradient-echo

          a) Nonspecific

       4) Enhancement with Gd-DTPA

          a) Slight and heterogeneous

5. Rhabdomyoma

    a. Most common cardiac tumor of infants and children

    b. Associated with tuberous sclerosis in 50%

    c. Almost always involve the myocardium

       1) Ventricles involved in 70%

          a) RV and LV equally involved

       2) Atria involved in 30%

       3) Multiple sites involved in 90% of cases

    d. Obstruction of cardiac chamber or valve present in 50%

    e. MR findings

       1) T1-weighted spin echo

          a) Homogeneous, slightly lower signal intensity than myocardium

       2) T2-weighted spin echo

          a) Increased signal intensity compared to myocardium

       3) Cine gradient-echo

          a) Very low signal intensity compared to myocardium

       4) Enhancement with Gd-DTPA

          a) Nonspecific

C. Malignant cardiac tumors

1. 25% of cardiac tumors

2. Metastases are 20-40 times more common than primary malignant cardiac tumors

3. Primary malignant cardiac tumors

    a. Most are sarcomas

       1) Angiosarcoma — most common (33%)

a) Usually located in RA
b) Arises from atrial septum
c) Typically presents with signs of right heart failure
d) High incidence of spread to pericardium with hemorrhagic pericardial effusion and pericardial tamponade.
e) MR findings
   i) High signal intensity on T1-weighted images due to areas of hemorrhage.
   ii) No significant change on T2-weighted images
   iii) Strong but often patchy signal enhancement with Gd-DTPA

2) Rhabdomyosarcoma
   a) Most common primary cardiac malignancy in infants and children; only accounts for 4-7% of all primary cardiac malignancies
   b) Multiple in 60%
   c) Most common sarcoma to arise from or involve a valve
   d) Often grows into a cardiac chamber markedly reducing the volume of the ventricular cavity
   e) Pericardial invasion common; may have associated hemorrhagic pericardial effusion
   f) MR findings
      i) Medium signal intensity similar to surrounding myocardium on T1-weighted images
      ii) Definite enhancement with Gd-DTPA

3) Leiomyosarcoma
   a) Usually arises from IVC
   b) Can arise from SVC
   c) Often grows into the RA
   d) MR shows signal intensity on T1-weighted images similar to liver but less than mediastinal fat.
   e) Often invades pericardium causing hemorrhagic pericardial effusion

4) Fibrosarcoma
   a) Diffusely infiltrates the myocardium often obliterating one or more cardiac chambers
   b) Often presents with signs and symptoms of congestive heart failure because of involvement of the LA
   c) MR findings
      i) Often isointense to myocardium on T1-weighted images
      ii) Increased signal intensity on T2-weighted images
      iii) Definite enhancement with Gd-DTPA

5) Liposarcoma
   a) Frequently arises in the pericardium and may invade the epicardium
   b) Has high, but heterogeneous signal intensity similar to subcutaneous fat on T1-weighted images

          c) Reduced or eliminated signal on fat-saturation images

          d) Slight enhancement with Gd-DTPA

     b. Pheochromocytoma

        1) Rare primary cardiac tumor

        2) Slightly decreased intensity or isointense to myocardium on T1-weighted images

        3) Very high signal intensity on T2-weighted images

     c. MR may not be able to provide a histologically specific diagnosis, but it is extremely useful in determining resectability and planning the surgical approach to the lesion.

     d. Lymphoma

        1) The heart is involved in up to 25% of cases of lymphoma but in vivo diagnosis is rare.

        2) Is slightly decreased in intensity or isointense to myocardium on T1-weighted images

        3) Isointense to myocardium on T2-weighted images

        4) Heterogeneous enhancement with Gd-DTPA with less enhancing central areas due to necrosis

4. Metastases

     a. Are much more common than primary cardiac tumors

     b. Occur in up to 20% of cases of malignant disease

     c. Hematogenous or lymphatic spread

        1) Bronchogenic carcinoma

        2) Breast carcinoma

        3) Melanoma

        4) Sarcoma

        5) Lymphatic malignancies including lymphoma and leukemia

     d. Direct invasion from adjacent thoracic malignancy

        1) Bronchogenic carcinoma

        2) Breast carcinoma

        3) Mediastinal lymphoma and other mediastinal malignancies

     e. Transvenous invasion

        1) Transvenous spread to the RA

          a) Renal cell carcinoma and occasionally hepatocellular carcinoma can spread to the RA via extension through the IVC.

          b) Bronchogenic carcinoma can occasionally spread to the RA via growth through the SVC

        2) Transvenous spread to the LA

          a) Bronchogenic carcinoma, primary pulmonary sarcoma, and occasionally metastatic tumors to the lung including metastatic chondrosarcoma, choriocarcinoma, and cervical carcinoma may invade the pulmonary veins and extend to the LA.

          b) Can produce a mobile intraatrial mass that can obstruct the mitral valve and mimic LA myxoma

c) Can be differentiated from LA myxoma by demonstrating anatomic continuity of the intraatrial mass and a mass in the lung via one of the pulmonary veins
   i) The invaded pulmonary vein is usually enlarged.
   ii) If the lung mass abuts the LA at the point of entry of one of the pulmonary veins that pulmonary vein may be totally obliterated.

VII. Myocardial diseases
A. Arrhythmogenic right ventricular dysplasia (ARVD)
1. Also known as arrhythmogenic right ventricular cardiomyopathy
2. Most common cause of arrhythmias originating from the RV
3. Familial disorder
   a. Autosomal dominant inheritance
   b. 30-50% of patients have a family history of sudden death at an early age, often <35 years of age
4. ARVD is characterized by thinning of the RV myocardium, which is also partially or completely replaced by adipose tissue, dilatation of the RV and RVOT, and RV wall motion abnormalities. Occasionally fatty replacement of the myocardium may also involve the interventricular septum and LV free wall.
5. MR is the best imaging modality for demonstrating the fatty replacement of the RV myocardium. The use of surface coils enhances the visualization of fat in the RV myocardium.
6. Electron beam CT and helical CT can also demonstrate the fat deposits in the RV myocardium and other features of ARVD.

B. Uhl's anomaly (parchment RV)
1. Rare anomaly characterized by severe hypoplasia or aplasia of the RV myocardium
   a. The endocardium and epicardium of the RV free wall are adjacent to each other and the RV free wall is paper-thin.
   b. The interventricular septum is normal.
2. The RV is markedly dilated and RV contractility is minimal. There is dilatation of the tricuspid valve annulus, severe tricuspid regurgitation, and marked right atrial enlargement (RAE).

C. Cardiomyopathy
1. Hypertrophic cardiomyopathy
   a. Hypertrophic cardiomyopathy is characterized by pronounced LV myocardial hypertrophy with decreased volume of the LV cavity. The RV is usually spared.
      1) Common cause of sudden death in the young
      2) Histology shows myocardial cellular disorganization and myocardial fiber disarray with patchy areas of tissue necrosis.
      3) The distribution of myocardial hypertrophy is usually evaluated with T-1 weighted spin echo techniques

4) Because of associated tissue necrosis there is often some degree of signal enhancement after administration of Gd-DTPA.

b. There are 2 main forms

   1) Obstructive type

      a) There is usually uniform thickening of the LV wall and interventricular septum.

      b) Associated with systolic anterior motion of the anterior mitral leaflet, obstruction of the left ventricular outflow tract (LVOT), mitral regurgitation, and LAE

      c) The systolic anterior motion of the anterior mitral leaflet often contributes significantly to the obstruction of the LVOT.

      d) Cine gradient-echo sequences show anterior motion of the anterior mitral leaflet during systole, narrowing of the LVOT (subaortic stenosis), a flow void in the LVOT during systole due to the subaortic stenosis, and a flow void in the LA during systole due to mitral regurgitation.

      e) The flow void in the LVOT is usually easily demonstrated but requires a short TE (about 4 msec) for optimum visualization.

   2) Nonobstructive type

      a) May have diffuse thickening of the entire LV

      b) There are two subtypes

         i) Asymmetric septal hypertrophy (ASH), in which the thickening of the myocardium predominantly or exclusively involves the interventricular septum

         ii) Apical hypertrophy, in which the myocardial thickening is predominantly limited to the LV apex

2. Dilated cardiomyopathy

  a. Most common form of cardiomyopathy

  b. Histology shows diffuse interstitial fibrosis with a decreased number of myocytes. There is often some degree of inflammatory myocarditis.

  c. Can involve the RV, LV, or both, but the LV is usually more severely involved

  d. Characterized by ventricular dilatation with poor systolic contractility and a reduction in ejection fraction

  e. The myocardial wall may remain of normal thickness but is often thinned, particularly in the apex and along the posteroinferior wall.

  f. Spin echo and cine gradient-echo sequences are useful in demonstrating the various features of dilated cardiomyopathy.

3. Restrictive cardiomyopathy

  a. Least common form

  b. Infiltration of myocardium by fibrosis or other tissue

  c. Characterized by severe diastolic dysfunction, biatrial enlargement, mitral regurgitation, and normal LV size and systolic function

  d. Atrial thrombi are common.

e. The major role of MR is excluding constrictive pericarditis by demonstrating normal pericardium in a patient with restrictive physiology.

f. In some infiltrating cardiac diseases, MR may be useful in documenting the presence of cardiac involvement.

  1) Sarcoidosis

    a) In systemic sarcoidosis the heart is involved in 20-30% of cases.

    b) Up to 50% of deaths from sarcoidosis may be related to cardiac involvement with congestive heart failure or sudden death.

    c) Patchy involvement of myocardium with granulomas which show high signal intensity on T1 and T2-weighted images

      i) The granulomas may have diffuse increased signal intensity.

      ii) They may also show central low-intensity signal with a high-intensity peripheral ring.

    d) Usually enhances with Gd-DTPA

    e) Because cardiac involvement is patchy, MR may be useful in deciding where to perform endomyocardial biopsy.

  2) Amyloidosis

    a) Usually has diffuse infiltration of myocardium with loss of atrial and LV function and left heart failure

    b) Usually enhances with Gd-DTPA

  3) Hemochromatosis

    a) Characterized by focal or diffuse iron deposition in the myocardium

    b) Results in myocardial thickening, ventricular dilatation with loss of function leading to congestive heart failure and death.

    c) Because of paramagnetic properties of iron, iron deposits cause loss of signal on T1-weighted, T2-weighted, and T2*-weighted images.

    d) The finding of patchy or diffuse loss of signal in dysfunctional myocardium coupled with a dark liver should allow a correct diagnosis.

  4) Endocardial fibroelastosis

    a) MR demonstrates thickening of the myocardium of the apices of both the RV and LV, which appear "V-shaped" on long axis views.

    b) Stroke volume is decreased.

    c) Apical thrombus formation is common.

    d) The RA and LA are usually enlarged.

VIII. Ischemic heart disease
  A. Assessment of cardiac function
    1. Both RV and LV volumes and mass can be accurately measured on short-axis cine gradient-echo images using the Simpson method previously described.
    2. Normal values in adults for LV volumes and mass
      a. In men (values quoted are mean ±1 standard deviation, with the 95% confidence interval for the normal range in brackets)
        1) LV end-diastolic volume — 136 ± 30 [77-195] ml
        2) LV end-systolic volume — 45 ± 14 [19-72] ml
        3) LV stroke volume — 92 ± 21 [51-133] ml
        4) LV ejection fraction — 67 ± 5 [56-78] %
        5) LV mass — 178 ± 31 [118-238] gm
      b. In women
        1) LV end-diastolic volume — 96 ± 23 [52-141] ml
        2) LV end-systolic volume — 32 ± 9 [13-51] ml
        3) LV stroke volume — 65 ± 16 [33-97] ml
        4) LV ejection fraction — 67 ± 5 [56-78] %
        5) LV mass — 125 ± 26 [75-175] gm
    3. Normal values in adults for RV volumes and mass
      a. In men
        1) RV end-diastolic volume — 157 ± 35 [88-227] ml
        2) RV end-systolic volume — 63 ± 20 [23-103] ml
        3) RV stroke volume — 95 ± 22 [52-138] ml
        4) RV ejection fraction — 60 ± 7 [47-74] %
        5) RV free wall mass — 50 ± 10 [30-70] gm
      b. In women
        1) RV end-diastolic volume — 106 ± 24 [58-154] ml
        2) RV end-systolic volume — 40 ± 14 [12-68] ml
        3) RV stroke volume — 66 ± 16 [35-98] ml
        4) RV ejection fraction — 63 ± 8 [47-80] %
        5) RV free wall mass — 40 ± 8 [24-55] gm
    4. Stroke volume is determined by subtracting the end-systolic volume from the end-diastolic volume.
    5. Ejection fraction is defined as the stroke volume divided by the end-diastolic volume.
  B. Assessment of wall motion
    1. Cine gradient-echo sequences coupled with pharmacologic stress can be used to detect reversible myocardial ischemia in patients with coronary artery disease.
    2. Dobutamine infusion, in doses up to 20 µg/kg/min, can be used to elicit myocardial wall motion abnormalities in patients with reversible ischemia.
    3. Dobutamine stress MR has 83-91% sensitivity and 80-100% specificity for detection of coronary artery disease.

   C. Myocardial perfusion imaging

     1. Fast gradient-echo sequences or echoplanar imaging (EPI), coupled with pharmacologic-induced stress and administration of Gd-DTPA can be used to demonstrate abnormalities in myocardial perfusion.

     2. Pharmacologic stress can be induced with adenosine, dipyridamole, or dobutamine. Dobutamine is preferred if myocardial wall motion is also being assessed. If not, either adenosine or dipyridamole can be used.

     3. First-pass, wash-in, and washout imaging following a bolus of Gd-DTPA is performed at rest. Following a brief period for equilibrium of the first Gd-DTPA dose, infusion of the stress agent is started and first-pass, wash-in, and washout imaging is again performed following a second bolus of Gd-DTPA.

     4. Underperfused myocardium appears as an area of decreased signal intensity, compared to surrounding normal myocardium that appears after the induction of pharmacologic stress.

     5. Areas of myocardial ischemia are usually more pronounced in the subendocardial myocardium than in the subepicardial myocardium. This is due to the greater metabolic demands and higher vascular resistance in the subendocardium compared to the subepicardium.

   D. Myocardial infarction (MI)

     1. Acute MI

       a. Spin echo sequences can be used to evaluate myocardial necrosis following an acute MI.

         1) Acute myocardial necrosis causes myocardial edema, which results in increased signal intensity on T2-weighted sequences.

         2) The area of myocardial necrosis will usually also demonstrate signal enhancement following administration of Gd-DTPA.

       b. Gradient-echo techniques coupled with administration of Gd-DTPA can also be used to evaluate myocardial necrosis following an acute MI.

         1) Following an acute MI, blood flow is often restored to the infarcted myocardium. Reperfusion of the infarcted area is often greater in the periphery of the infarct. Sequestration of neutrophils in the microvessels in the center of the infarct can lead to microvessel occlusion by erythrocytes, leukocytes, and cellular debris preventing reperfusion of the infarct core.

         2) If the central core of the MI is not reperfused it will appear as an area of low signal intensity following administration of Gd-DTPA.

         3) Reperfused, infarcted myocardium appears as a hyperenhanced area of myocardium on delayed imaging (>10 minutes after contrast administration). This allows accurate assessment of infarct size.

     2. Myocardial viability

       a. MR signs of myocardial viability following an acute MI include: absence of increased signal intensity on spin echo images of the infarcted area of myocardium, any sign of wall thickening at rest, wall-thickening after stimulation by low-dose dobutamine, and preserved wall thickness.

b. Signs of nonviability include high signal intensity on spin echo images, enhancement of the infarct following administration of Gd-DTPA (possibly with a low signal central core), reduced wall thickness, and absence of wall thickening in response to dobutamine.

3. Chronic MI
   a. Usually appears as an area of end-diastolic myocardial thinning with absent systolic thickening or systolic thinning.
   b. Cine gradient-echo sequences may also demonstrate decreased, absent, or paradoxical motion of the involved portion of the ventricular wall during systole.

4. Complications of MI
   a. False LV aneurysm — cine gradient-echo sequences show a focal disruption of the posterior LV myocardium with high signal intensity blood in the LV cavity that communicates with a large retrocardiac collection of blood via the disrupted LV wall
   b. True LV aneurysm — seen as enlargement of the cavity of the LV apex associated with thinning of the LV apical myocardium, which demonstrates reduced, absent, or paradoxical motion during systole
   c. Acquired ventricular septal defect (VSD) — cine gradient-echo sequences can show the defect in the interventricular septum and a flow void in the RV from the turbulent jet of flow through the VSD
   d. Papillary muscle dysfunction with mitral regurgitation — cine gradient-echo sequences can show MI involving the papillary muscles and a flow void in the LA above the mitral valve during ventricular systole.

E. Coronary MR angiography (MRA)
   1. Navigator echoes are usually used to minimize the effects of respiratory motion. Free-breathing techniques are preferred over multiple breath-hold techniques because they require less patient cooperation.
   2. Coronary MRA can be performed by several methods.
      a. Black blood techniques
         1) Spin echo
            a) Conventional ECG-gated spin echo coronary MRA was able to identify the coronary ostia and occasionally portions of the coronary arteries.
            b) Has been abandoned because of inability to detect anomalous coronary artery anatomy or coronary atherosclerosis
         2) Dual inversion fast spin echo
            a) ECG-gated, navigator free-breathing dual inversion fast 2-D spin echo allows performance of high-quality coronary MRA.
            b) The timing of the inversion pulse is chosen to null signal from the coronary blood pool.
            c) White blood techniques have problems accurately delineating luminal stenosis because of signal voids produced by turbulent blood flow. Furthermore, high signal thrombus, vessel wall, and

certain plaque components may obscure coronary stenosis on white blood coronary MRA. Black blood fast spin echo coronary MRA does not have these problems and may be better at identifying coronary artery disease.

d) In addition, artifacts due to local magnetic field distortion from metallic implants such as vascular clips are a problem with white blood techniques since they accentuate these artifacts. Black blood techniques minimize these artifacts and are particularly advantageous for patients with vascular clips and other metallic implants.

b. White blood techniques
1) Segmented k-space gradient-echo technique
a) 2-D breath-hold technique
i) Used for evaluation of coronary arteries and coronary artery bypass grafts
ii) Artifacts from graft markers and vascular clips hinder evaluation of patients with coronary artery bypass grafts.
b) 3-D navigator free-breathing technique
i) Has superior signal-to-noise ratio and post-processing capabilities
ii) Currently the most widely used technique
2) Segmented EPI
a) Because of the speed of EPI, acquisition times are greatly reduced.
b) The segmentation of k-space reduces artifacts due to blood flow and motion.
3) Spiral acquisition
a) Spiral acquisitions have more efficient filling of k-space, enhanced signal-to-noise ratio, and favorable flow properties.
b) Implementation has been limited because of the complexity of reconstruction algorithms.
4) Contrast-enhanced coronary MRA
a) White blood techniques are dependent upon the inflow of blood into the imaging plane. If there is slow flow, saturation effects will cause loss of signal. Furthermore, vessel wall, plaque, and thrombus can have similar signal intensity to blood.
b) Contrast-enhanced coronary MRA could allow for true imaging of the lumen.
c) Gd-DTPA is not ideal for coronary MRA because of rapid vascular equilibration and extravasation into the extravascular space, but it can be used for first-pass studies. Specific intravascular contrast agents for coronary MRA are under development.
5) For white blood coronary MRA techniques, fat-saturation is used to eliminate the signal from surrounding epicardial fat and enhance the signal from coronary blood flow.

3. Coronary MRA can demonstrate areas of atherosclerotic narrowing and occlusion.
4. Coronary MRA can also be used to demonstrate congenital anomalies of coronary artery anatomy, to evaluate coronary artery patency after angioplasty, and to evaluate patency of coronary artery bypass grafts.
5. Coronary artery stents pose a particular problem for coronary MRA.
   a. Coronary artery stents are made from high-grade stainless steel, tantalum, or alloy.
   b. Although the attractive force and local heating are negligible at 1.5 T, artifacts due to local magnetic field distortion from these metallic stents can be significant. The resulting signal void precludes direct visualization of the lumen within the stent and evaluation of coronary artery patency adjacent to the stent.
   c. Evaluation of blood flow proximal and distal to the stent may provide indirect evidence of patency.
   d. Should wait until 8 weeks following stenting to do MR

IX. Congenital heart disease
A. Great vessel abnormalities
1. Coarctation of the aorta
   a. Most often there is narrowing of the aorta just distal to the left subclavian artery and just distal to the ductus arteriosus.
   b. Collateral circulation comes from the costocervical and thyrocervical branches of the subclavian artery, and the internal mammary, and intercostal arteries.
   c. 50% have a bicuspid aortic valve; coarctation is also the most common cardiovascular anomaly found in Turner's syndrome
   d. Spin echo sequences are useful in demonstrating the site and extent of narrowing of the aorta, the poststenotic dilatation, and the collateral circulation. Cine gradient-echo sequences can also demonstrate the jet of turbulent flow through the coarcted segment and can demonstrate an associated bicuspid aortic valve.
2. Aberrant right subclavian artery
   a. A left aortic arch with an aberrant right subclavian artery is the most common congenital anomaly of the aortic arch.
   b. The order of branching is as follows: right common carotid artery, left common carotid artery, left subclavian artery, right subclavian artery.
   c. The aberrant right subclavian artery may arise directly from the distal portion of the left aortic arch or it may arise from an aortic diverticulum (diverticulum of Kommerell) that represents persistence of the most distal portion of the embryologic right arch.
   d. Forms a loose vascular ring — respiratory symptoms and dysphagia may occur but are uncommon
   e. Spin echo or cine gradient-echo images typically demonstrate well the aberrant vessel in its entire course.

3. Right aortic arch
    a. Two major types
        1) Mirror-image branching
            a) The order of branching is as follows: left innominate artery, right common carotid artery, and right subclavian artery.
            b) There is no aortic diverticulum.
            c) Does not produce a vascular ring
            d) Has a high association with cyanotic cardiac defects
                i) Truncus arteriosus — 40% have mirror-image right aortic arch
                ii) Tetralogy of Fallot (TOF) — 25-30% have mirror-image right aortic arch
                iii) Tricuspid atresia — 5-10% have mirror-image right aortic arch
                iv) Transposition of the great vessels (TGV) — 3% have mirror-image right aortic arch
        2) Right aortic arch with aberrant left subclavian artery
            a) The left subclavian artery, rather than arising from a left innominate artery, arises separately as the last major branch of the aortic arch. The order of branching is as follows: left common carotid artery, right common carotid artery, right subclavian artery, left subclavian artery.
            b) The aberrant left subclavian artery may arise directly from the distal right aortic arch, but usually arises from an aortic diverticulum.
            c) Forms a loose vascular ring
                i) Respiratory symptoms and dysphagia may occur but are uncommon.
                ii) Still, right aortic arch with aberrant left subclavian artery is the second most common cause of a symptomatic vascular ring requiring surgery.
            d) In most instances the anomaly is an isolated one, although 5% have associated cardiac defects that are usually of the acyanotic type.
    b. MR is useful in demonstrating the right aortic arch and in distinguishing between the two major types.
4. Double aortic arch
    a. The right arch is usually larger and slightly higher in position and courses posterior to the trachea and esophagus. The left arch courses anterior to the trachea and esophagus.
    b. Forms a tight vascular ring
        1) Respiratory symptoms and dysphagia are common.
        2) Most common cause of a symptomatic vascular ring requiring surgery

c. MR is useful in demonstrating the two arches and in showing the right and left common carotid and subclavian arteries arising from their respective arches.

5. Left pulmonary artery sling
   a. Rare anomaly
   b. The left pulmonary artery arises from the proximal portion of the right pulmonary artery and passes through the mediastinum between the trachea and esophagus to supply the left lung.
   c. Can compress both the trachea and esophagus
   d. MR is useful in demonstrating the abnormal origin and course of the left pulmonary artery.

B. Valvular pulmonic stenosis
   1. Usually a congenital anomaly
   2. The leaflets are fused at the commissures preventing them from opening completely in systole.
   3. In valvular pulmonic stenosis, cine gradient-echo sequences show valve leaflet thickening and bulging with a systolic flow jet in the MPA. There is also usually RV wall thickening or RV enlargement.

C. Left-to-right shunts
   1. Atrial septal defect (ASD)
      a. Classification is based on location of the defect
         1) 70% — ostium secundum or fossa ovalis type ASD — centrally located in the septum and typically large
         2) 20% — ostium primum ASD — located just above the atrioventricular (AV) valves
         3) 6% — sinus venosus ASD — located adjacent to the entrance of the SVC into the RA and appears as a defect between the posterior border of the SVC and the LA
      b. The defect in the atrial septum is usually well shown by spin echo images.
      c. Cine gradient-echo images can also confirm the left-to-right shunt by showing a fan-shaped flow void in the RA near the septal defect.
   2. VSD
      a. Classification is based on location of the defect
         1) 80% — membranous VSD — defect in the superior aspect of the interventricular septum immediately below the aortic valve and adjacent to the septal leaflet of the tricuspid valve
         2) 10% — muscular VSD — defect in the muscular portion of the interventricular septum
         3) 5-10% — posterior VSD — posterior defect in the membranous septum extending below the tricuspid valve associated with AV canal
         4) 5% — supracristal VSD — defect in the interventricular septum immediately below the pulmonic valve
      b. Most VSDs are well shown by spin echo images.

      c. Muscular VSD may be difficult to image on spin echo sequences; however, the diagnosis can be made by demonstration of a flow void in the RV adjacent to the muscular portion of the interventricular septum on cine gradient-echo images.

  3. AV canal (endocardial cushion defect)

      a. Most common cardiac anomaly in patients with Down's syndrome

      b. Endocardial cushion defects result in a broad spectrum of cardiac abnormalities ranging from an isolated cleft in an AV valve to a complex anomaly that includes an ostium primum ASD, posterior membranous VSD, and mitral and tricuspid insufficiency.

      c. Exact classification by MR is difficult, but MR can show the common AV valve and associated ASD and VSD, and cine gradient-echo sequences can demonstrate the left-to-right shunt.

  4. Patent ductus arteriosus (PDA)

      a. Both spin echo and cine gradient-echo sequences can demonstrate the PDA coursing between the inferior aspect of the aortic arch just distal to the origin of the left subclavian artery and the top of the left pulmonary artery just distal to its origin.

D. Cyanotic lesions

  1. TOF

      a. Represents 8% of congenital heart disease

      b. Components consist of VSD; pulmonic stenosis, which is usually infundibular; an overriding aorta which straddles the VSD; and RV hypertrophy

      c. 25-30% also have a right aortic arch with mirror-image branching

      d. MR is useful in demonstrating these components of TOF and can be used to evaluate the status of systemic-to-pulmonary artery collaterals and palliative surgical shunts.

  2. Tricuspid atresia

      a. The anterior atrioventricular ring is replaced by fat and there is no continuity between the RA and RV. Spin echo MR shows a broad band of fat where the tricuspid valve should be.

      b. The RV is usually hypoplastic.

      c. The RA is enlarged and there is usually an ASD. Blood typically reaches the lungs via a VSD; however, if the ventricular septum is intact pulmonary blood flow is totally dependent upon a PDA.

      d. Pulmonic stenosis or atresia and TGV may also coexist.

  3. Ebstein's anomaly

      a. There is downward displacement of the tricuspid valve into the RV with an enlarged RA, "atrialized" RV, and a small trabecular RV chamber.

      b. The anterior leaflet of the tricuspid valve is normally attached at the atrioventricular junction, but distally it is abnormally attached to the RV wall.

      c. The posterior and septal leaflets of the tricuspid valve are small or absent.

  d. Cine gradient-echo sequences show very limited contraction of the RV and massive tricuspid regurgitation.

  e. ASD is common.

 4. D-TGV

  a. Characterized by ventriculoarterial discordance in the presence of atrioventricular concordance.

  b. In other words, the atria and ventricles are normally related. The LA sits to the left of the RA and the LV still sits to the left of the RV. The great vessels connect to the wrong ventricle.

  c. Systemic venous blood flows into the anatomic RA, and then into the RV, which supplies the aorta. Pulmonary venous flow returns to the LA, and then into the LV, which supplies the MPA.

  d. The systemic and pulmonary circulations are parallel — requires a shunt between the two to support life.

   1) VSD in 50%

   2) In 50% the ventricular septum is intact. Admixture of blood occurs through a PDA and a patent foramen ovale.

  e. MR is useful in demonstrating the components of D-TGV.

   1) The relationship between the MPA and ascending aorta is reversed.

    a) Normally the MPA is positioned to the left of the ascending aorta.

    b) In D-TGV the MPA will either be positioned directly behind the ascending aorta or to the right of the ascending aorta.

   2) The ascending aorta arises from the RV.

   3) Anatomic characteristics of the RV that allow it to be identified with certainty by MR include:

    a) The RV has a well-formed outflow tract seen as a thick-walled muscular structure resembling a doughnut in axial section.

    b) The RV is more trabeculated than the LV.

    c) The RV contains the moderator band, a thick muscular band crossing the RV cavity from the free wall to the interventricular septum near the RV apex.

   4) The MPA arises from the LV.

   5) 3% have mirror-image right aortic arch

 5. Congenitally corrected transposition (L-TGV)

  a. There is ventriculoarterial discordance in the presence of atrioventricular discordance.

  b. The atria are normally related; however, the ventricles are now connected to the wrong atrium and the great vessels are connected to the wrong ventricle. The LA still sits to the left of the RA, but the RV now sits to the left of the LV. The ascending aorta arises from the RV and the MPA arises from the LV.

  c. Systemic venous blood flows into the anatomic RA, then into the anatomic LV, and then out the MPA; pulmonary venous flow returns to the anatomic LA, then into the anatomic RV, and then out the aorta.

      d. Thus the abnormalities are congenitally "corrected."

      e. MR is useful in demonstrating the features of corrected transposition

          1) The relationship of the MPA and ascending aorta is still abnormal. Again the MPA will either sit behind or to the right of the ascending aorta.

          2) The ascending aorta arises from the RV; however, the RV now sits to the left of the LV and the RV, RVOT, and ascending aorta form the left heart border. This is well seen on coronal images.

          3) The MPA arises from the LV.

      f. In most cases of corrected transposition other abnormalities coexist and MR is useful in demonstrating these abnormalities too.

          1) Insufficiency of the left AV (anatomic tricuspid) valve, leading to features of "mitral" regurgitation

          2) VSD, usually large

          3) Pulmonic stenosis, usually associated with VSD, and presenting with a clinical picture resembling TOF

6. Truncus arteriosus

    a. Results from failure of formation of the spiral septum within the truncus in the early embryo

    b. As a result, a single arterial vessel (the truncus) leaves the heart.

    c. The pulmonary arteries arise from the truncus.

    d. The truncus sits atop a large VSD.

    e. 40% have a right aortic arch with mirror-image branching

    f. MR is useful in demonstrating these components of truncus arteriosus.

7. Totally anomalous pulmonary venous return

    a. The four pulmonary veins do not converge on the LA; instead, they converge on a "confluence" of pulmonary veins, known as the common pulmonary vein, which maintains a connection to the primitive systemic venous system. The common pulmonary vein sits directly behind the LA but does not communicate with the LA.

    b. Four forms

        1) Supracardiac connection to a right SVC — persistence of communication to the right anterior cardinal system

        2) Supracardiac connection to a left SVC or vertical vein — persistence of communication to the distal part of the left anterior cardinal system (produces the "snowman" or "figure-of-8" configuration)

        3) Cardiac connection to the coronary sinus or RA — persistence of communication to the proximal part of the left anterior cardinal system

        4) Infracardiac connection to the portal venous system—persistence of communication to the umbilical-vitelline system

    c. ASD is an integral component of the anomaly.

    d. MR can show the anomalous venous drainage and the ASD.

References and suggested additional reading

1. Alpert JS, Dalen JE, Rahimtoola SH. *Valvular Heart Disease 3rd ed.* Philadelphia: Lippincott Williams & Wilkins, 2000.
2. Boxt LM. How to perform cardiac MR imaging. MRI Clin North Am 1996; 4:191-216.
3. Boxt LM. MR imaging of acquired heart disease. MRI Clin North Am 1996; 4:253-268.
4. Boxt LM. MR imaging of congenital heart disease. MRI Clin North Am 1996; 4:327-359.
5. Cheriex EC, Pieters FAA, Janssen JHA, de Swart H, Palmans-Meulemans A. Value of exercise Doppler-echocardiography in patients with mitral stenosis. Int J Cardiol 1994; 45:219-226.
6. Didier D, Ratib O, Lerch R, Friedli B. Detection and quantification of valvular heart disease with dynamic cardiac MR imaging. RadioGraphics 2000; 20:1279-1299.
7. Duerinckx AJ. Coronary MR angiography. MRI Clin North Am 1996; 4:361-418.
8. Duerinckx AJ. Imaging of coronary artery disease—MR. J Thorac Imaging 2001; 16: 25-34.
9. Duerinckx AJ, Higgins CB, Pettigrew RI. *MRI of the Cardiovascular System.* New York: Raven, 1994.
10. Flamm SD, VanDyke CW, White RD. MR imaging of the thoracic aorta. MRI Clin North Am 1996; 4:217-235.
11. Fuster V, Alexander RW, O'Rourke RA, Roberts R, King SB III, Wellens HJJ. *Hurst's The Heart 10th ed.* New York: McGraw-Hill, 2001.
12. Gedgaudas E, Moller JH, Castaneda-Zuniga WR, Amplatz K. *Cardiovascular Radiology.* Philadelphia: Saunders, 1985.
13. Goldin JG, Ratib O, Aberle DR. Contemporary cardiac imaging an overview. J Thorac Imaging 2000; 15:218-229.
14. Greenberg SB. Assessment of cardiac function: magnetic resonance and computed tomography. J Thorac Imaging 2000; 15:243-251.
15. Hall RJC, Julian DG. *Diseases of the Cardiac Valves.* Edinburgh: Churchill Livingstone, 1989.
16. Kimura F, Sakai K, Sakomura Y, Kujimura M, Ueno E, Matsuda N, Kasanuki H, Mitsuhashi N. Helical CT features of arrhythmogenic right ventricular cardiomyopathy. RadioGraphics 2002; 22:1111-1124.
17. Manning WJ, Pennell DJ. *Cardiovascular Magnetic Resonance.* New York: Churchill Livingstone, 2002.
18. Miller SW. *Cardiac Radiology: The Requisites.* St. Louis: Mosby, 1996.
19. NessAiver M. A guide to cardiac MRI. http://www.simplyphysics.com, 2000.
20. Otto CM. *Valvular Heart Disease.* Philadelphia: Saunders, 1999.
21. Park JH, Kim YM. MR imaging of cardiomyopathy. MRI Clin North Am 1996; 4:269-286.

John H. Woodring, M.D.

22. Peshock RM, Willett DL, Sayad DE, Hundley WG, Chwialkowski MC, Clarke GD, Parkey RW. Quantitative MR imaging of the heart. MRI Clin North Am 1996; 4:287-305.
23. Reddy GP, Higgins CB, Chao KH, Tung PP. *Cardiac MR Imaging*. Philadelphia: Lippincott Williams & Wilkins, 2001.
24. Rozenshtein A, Boxt LM. Computed tomography and magnetic resonance imaging of patients with valvular heart disease. J Thorac Imaging 2000; 15:252-264.
25. Schmailzl KJG, Ormerod O. *Ultrasound in Cardiology*. Berlin: Blackwell, 1994.
26. Schvartzman PR, White RD. Imaging of cardiac and paracardiac masses. J Thorac Imaging 2000; 15:265-273.
27. Schwitter J, Sakuma H, Saeed M, Wendland MF, Higgins CB. Very fast cardiac imaging. MRI Clin North Am 1996; 4:419-432.
28. Spindola-Franco H, Fish BG. *Radiology of the Heart: Cardiac Imaging in Infants, Children, and Adults*. New York: Springer-Verlag, 1985.
29. White CS. MR evaluation of the pericardium and cardiac malignancies. MRI Clin North Am 1996; 4:237-251.
30. Woodring JH, Bognar B, van Wyk CS. Metastatic chondrosarcoma to the lung with extension into the left atrium via invasion of the pulmonary veins: presentation as embolic cerebral infarction. Clin Imaging 2002; 26:338-341.

Index

A

ABOUT THE AUTHOR:

John H. Woodring, M.D. has published over 100 scientific articles pertaining to cardiac and thoracic radiology. An expert teacher, he has been recognized four times as the "Outstanding Teacher of the Year" by his residents. He is a fellow of both the American College of Chest Physicians and the American College of Radiology, and is a staff radiologist at the Lexington Department of Veterans Affairs Medical Center and an Adjunct Professor of Diagnostic Radiology at the University of Kentucky Medical Center in Lexington, Kentucky. In this book, Dr. Woodring combines an approach to the interpretation of chest radiographs useful on oral board exams in cardiac radiology with basic material that is covered on the written board exams.

www.ingramcontent.com/pod-product-compliance
Lightning Source LLC
Chambersburg PA
CBHW081111170526
45165CB00008B/2411